INTRODUCTION TO QUANTUM THEORY

Since its emergence in the early twentieth century, quantum theory has become the fundamental physical paradigm, and is essential to our understanding of the world.

Providing a deeper understanding of the microscopic world through quantum theory, this supplementary text covers a wider range of topics than conventional textbooks. Emphasis is given to modern achievements such as entanglement, quantum teleportation, and Bose–Einstein condensation. Macroscopic quantum effects of practical relevance, for example, superconductivity and the quantum Hall effect, are also described. Looking to the future, the author discusses the exciting prospects for quantum computing.

Physical, rather than formal, explanations are given, and mathematical formalism is kept to a minimum so that readers can understand the concepts. Theoretical discussions are combined with a description of the corresponding experimental results. This book is ideal for undergraduate and graduate students in quantum theory and quantum optics.

HARRY PAUL is a retired Professor of Theoretical Physics at the Institute of Physics, Humboldt University Berlin. His research interests are in nuclear physics and quantum theory, especially laser theory, nonlinear optics and quantum optics.

INTRODUCTION TO
QUANTUM THEORY

HARRY PAUL

Humboldt-Universität zu Berlin

CAMBRIDGE
UNIVERSITY PRESS

CAMBRIDGE UNIVERSITY PRESS
Cambridge, New York, Melbourne, Madrid, Cape Town, Singapore, São Paulo, Delhi

Cambridge University Press
The Edinburgh Building, Cambridge CB2 8RU, UK

Published in the United States of America by Cambridge University Press, New York

www.cambridge.org
Information on this title: www.cambridge.org/9780521876933

First published 2008

Printed in the United Kingdom at the University Press, Cambridge

A catalogue record for this publication is available from the British Library

ISBN 978-0-521-87693-3 hardback

Contents

Preface

This is an unconventional introduction to quantum mechanics. Emphasis is laid on the physical aspects rather than the formal apparatus. (As a side-effect, the book will be easier to read than usual textbooks in which mathematics is predominant.) To elucidate the novel features displayed by the quantum world, comparison is made with classical physics whenever possible, thus emphasizing the anti-intuitive (in German: 'unanschaulich') character of the new physics.

It is my goal to discuss thoroughly the basic quantum mechanical concepts, such as quantum states and their preparation, quantum mechanical uncertainty, quantum correlations and quantum measurement. In addition, selected experiments are reported that show up the potential of quantum theory and, by the way, tell us of the ingenuity of the researchers. Many of those experiments are taken from quantum optics which, in fact, is a wonderful playing ground for quantum physicists. In particular, exciting theoretical concepts, such as the famous gedanken experiment of Einstein, Podolsky and Rosen, and quantum teleportation, could be realized experimentally. Those achievements, of course, find their due place in this book.

Attention is also given to the mysterious interrelationship between spin and statistics, which makes the behaviour of Fermi and Bose gases so different. A special paragraph is devoted to the experimental verification of Bose–Einstein condensation, one of the highlights of experimental research in the last two decades.

Furthermore, fundamental interaction processes, notably scattering, are treated, as well as macroscopic quantum effects, such as superconductivity and the Josephson and quantum Hall effects. Finally, the intriguing concept of quantum computers is briefly discussed.

With this book, I hope to contribute to a deeper understanding of quantum mechanics and its 'mysteries'. So it might supplement, certainly not substitute, existing textbooks on quantum theory and quantum optics. It addresses mainly students and scientific workers. However, they should already be familiar with the quantum mechanical formalism.

What fascinates me on quantum physics is, first of all, ideas. I hope, I can convey some of my enthusiasm to the reader.

1

Unexpected findings

1.1 Prelude: a new constant enters the stage

One year before his epoch-making work on blackbody radiation in which he
postulated the existence of an elementary quantum of action, Max Planck (1899)
proposed a system of natural units that was based on four universal constants. They
were the velocity of light c, the gravitational constant f, and two new constants
a, b that he took from Wien's radiation law (which he, at the time, believed to be
strictly valid). He determined the numerical values of a and b from measurements
of both the total energy emitted from a blackbody source and the constant in
the exponent of Wien's law. In retrospect, since we know that Wien's law is
a good approximation to the true radiation law (found by Planck in 1900), it
becomes clear that b is nothing but the famous elementary quantum of action
known as Planck's constant h, while a is identical to the ratio of h and Boltzmann's
constant k.

Requiring all four constants to take the numerical value unity, Planck arrived
at a system of physical units which he was quite enthusiastic about. He wrote
(Planck, 1899): 'It might not be without interest to remark that with the aid of
the two ... constants a and b the possibility is given to establish units for length,
mass, time and temperature, which, independently of special bodies or substances,
necessarily retain their importance for all times and for all, even extraterrestric
and extrahuman cultures and which therefore can be termed 'natural measurement
units'.[1] However, at the time little (if any) attention was given to his proposal.
This is no wonder since the units of length ($\approx 10^{-33}$ cm) and time ($\approx 10^{-43}$s) are
ridiculously small, whereas, on the contrary, the unit of mass ($\approx 10^{-5}$g) is moderate
and, finally, the unit of temperature ($\approx 10^{32}$ °C) is crazily large. Actually, however,

---

[1] Es nicht ohne Interesse sein zu bemerken, dass mit Zuhülfenahme der beiden ... Constanten a und b die
Möglichkeit gegeben ist, Einheiten für Länge, Masse, Zeit und Temperatur aufzustellen, welche, unabhängig
von speciellen Körpern oder Substanzen, ihre Bedeutung für alle Zeiten und für alle, auch ausserirdische und
aussermenschliche Culturen nothwendig behalten und welche daher als „natürliche Maasseinheiten" bezeichnet
werden können.

the physical relevance of Planck's units was revealed only by modern cosmology. Anyway, it is interesting to see that Planck had already smuggled the quantum of action into physics one year before his great achievement, the derivation of his famous radiation law.

Curiously enough, the detection of the elementary quantum of action and its intimate connection with energy quantization was the culmination of a long-standing search in *macroscopic* thermodynamics. The problem, posed as early as 1859, was to derive the spectral distribution function for blackbody radiation. This proved to be a real challenge. In 1913 Einstein commented on the pains that were taken in achieving this goal with the words: 'It would be uplifting if we could put on scales the substance of brain that was sacrificed by the theoretical physicists on the altar of this universal function.'[2] Neither Planck nor any other theorist desired to explore the microphysical world. Nevertheless, with his assumption that an oscillator, having the resonance frequency v, in the presence of an electromagnetic field exchanges energy with it only in portions of hv, he laid the foundation-stone to what eventually became the marvellous edifice of quantum mechanics. However, at the time neither he nor his contemporaries were aware of the revolutionary character of his quantization hypothesis. Even Einstein, in his famous 1905 paper on photons (Einstein, 1905), did not mention Planck's work. (So you will miss Planck's constant in Einstein's paper; actually, he used instead an equivalent combination of the universal gas constant, Avogadro's number and a constant taken from Wien's radiation law.)

In 1931 Planck confessed in a letter to Robert Williams Wood: 'This was a purely formal assumption, and I did not think much about it, instead I only thought that I had to bring about a positive result in any case, whatever it might cost.'[3] Nowadays we know that Planck's constant plays *the* decisive role in the microcosm. In fact, it gives us the scale for all quantum phenomena. (Some theorists like to consider classical physics as the limiting case of quantum theory for $h \rightarrow 0$.)

However, we should not forget the great share experimentalists had in Planck's success. The first achievement was the experimental realization by Wien, jointly with Lummer, of a reliable source emitting blackbody radiation, in the form of a heated cavity ('Hohlraum') with a small hole. Thus it became possible, for the first time, to investigate blackbody radiation thoroughly. A second experimental progress was Rubens' reststrahlen method, which allowed the examination of the long-wavelength part of the cavity radiation. Careful radiation measurements by

---

[2] Es wäre erhebend, wenn wir die Gehirnsubstanz auf die Waage legen könnten, die von den theoretischen Physikern auf dem Altar dieser universellen Funktion hingeopfert wurde.

[3] Das war eine rein formale Annahme, und ich dachte mir nicht viel dabei, sondern eben nur das, dass ich unter allen Umständen, koste es, was es wolle, ein positives Resultat herbeiführen mußte.

various workers at the Physikalisch-Technische Reichsanstalt at Berlin eventually indicated serious deviations from Wien's radiation formula that lay definitely outside the limits of error. Finally, when Rubens and Kurlbaum observed that the spectral distribution function in the long-wavelength infrared region showed a linear increase with temperature T rather than an exponential dependence on $-1/T$ as predicted by Wien's law, Planck could not but admit the failure of the latter. Actually, he felt under pressure to derive an improved radiation law that was in accordance with the experimental findings.

1.2 Observing the invisible

At first glance, it seems impossible to get reliable information on single microscopic systems. In fact, any kind of observation requires a *macroscopic* measuring apparatus. (Theorists like to think of pointer positions indicating the outcome of the measurement. Nowadays, however, the measuring results are fed as electric signals directly into computers for further evaluation. With this procedure, an old problem in the interpretation of quantum mechanics becomes obsolete: no human observer – with his consciousness – is needed to take notice of the measurement result.) Microscopic systems, however, cannot affect a macroscopic system directly such as to change its (macroscopic) state. This is simply because a microscopic system can transfer only a tiny energy to any other one.

Fortunately, this is not the last word. Actually, there are several mechanisms that allow us to enhance a primary microscopic signal so that it becomes macroscopic. In the following, I will shortly describe some important ones.

1.2.1 Multiplication by impact ionization

A charged particle, such as an electron or an ion, can readily be accelerated by a static electric field. Having thus gained sufficient kinetic energy, the particle is capable of producing electron–ion pairs in collisions with gas particles, or releasing secondary electrons when hitting the surface of a solid. Those newly generated charged particles become accelerated too and produce a third generation of charged particles. This process can be repeated several times, resulting in an avalanche of charged particles, which is readily transformed into a (macroscopic) electric signal. The latter gives us the information that a charged particle *has been* present shortly before in a certain region of space.

The present scheme was successfully applied first in the counter tube devised for the detection of charged particles, and also γ quanta, emitted in radioactive decay processes. The so-called Geiger–Müller counter, or Geiger counter, consists of a cylindrical tube (cathode) with a wire (anode) on its axis. The tube is filled with helium, and the operating voltage is rather high (about 1 kV) so that a single primary particle eventually gives rise to a self-sustaining electric discharge.

Of great practical relevance is the case that the primary particle, an electron, is generated, via the photoelectric effect, by an incident photon. Making this electron the ancestor of an electron avalanche enables us to detect the photon. In a photomultiplier, the primary electron becomes accelerated by an electric field and is directed to a metallic surface, a so-called dynode. There it produces, via impact ionization, several secondary electrons which, in turn, generate new electrons on a second dynode. This game is repeated several times, yielding, finally, a pulse of millions of electrons. In an electric circuit, this electron avalanche gives us an electric signal (a current or a voltage pulse) indicating that a photon has arrived a short time before on the detector's sensitive surface.

High spatial resolution in photodetection is achieved by using a microchannel plate for secondary electron production. This device is a glass disc with a large number of microchannels in it. Both sides of the disc are coated with metallic electrodes to which a voltage of some kV is applied. As a result, any electron entering one of the channels becomes accelerated. Since the channels are slightly tilted to the surface normal, the electron will hit the channel wall, thereby producing several secondary electrons. Through subsequent collisions with the wall, more and more electrons are generated.

A photomultiplier can also be operated in a continuous regime. Then it is a very efficient instrument to detect weak-intensity fields, whereby the photocurrent follows temporal intensity variations on timescales exceeding the detector's response time.

1.2.2 Transition from unstable to stable equilibrium

It is well known in classical physics that unstable equilibrium is extremely sensitive to perturbations. Think, for instance, of a ball that rests on the top of a stick. A slight kick will make it fall down, thus producing a large effect. Similarly, unstable thermodynamic states can be used for quantum measurements. In Wilson's cloud chamber, oversaturated steam is generated through a sudden expansion. A fast charged particle flying through the chamber ionizes the steam along its path. The ions thus produced act as condensation nuclei for water droplets. In this way, the path of a particle becomes visible, under suitable illumination, as a condensation track.

In a bubble chamber, on the other hand, a liquid is put into a superheated state through suddenly lowering its pressure. In such a medium, a charged particle of high energy makes the liquid boil along its path. As a result, visible bubbles are formed that indicate the path.

1.2.3 Multiplication by chemical processes

Actually, this is what happens in photography. The primary process takes place in a silver bromide grain, which is an ionic crystal built of Ag^+ and Br^- ions. An incident photon releases a valence electron from a bromine ion (inner photoelectric effect). The electron is freely movable within the lattice and can be captured by a crystal defect. As a result, the latter becomes negatively charged, thus being able to attract and neutralize a silver ion sitting at an interstitial site. This silver ion serves as a development nucleus: development causes frequent repetitions of the former process at the same defect, with the only difference that the electron is now delivered by the developing agent rather than released through photon absorption. In this way, the nucleus grows more and more. Having reached a critical size, a new mechanism – an electrolytic process – comes into play, which reduces the *whole grain* to metallic silver, thus producing a blackened spot on the photographic film.

A masterpiece in light detection is certainly the eye developed by nature over a long period of time. The mechanism underlying vision is, in fact, a complicated, basically chemical process. The primary process is the absorption of a photon in a rhodopsin molecule located in a retinal rod, which leads, via several isomeric states, to an activated rhodopsin molecule R^* (metarhodopsin II). The R^* molecule, in turn, starts a biochemical cascade in which several enzymatic reactions cooperate, the result being the hydrolysis of about 1000 cyclic guanosine-monophosphate (cGMP) molecules. Then the R^* molecule initiates the same cascade again, and this procedure is repeated about 100 times. In this way, a single photon gives rise to the hydrolysis of about 100 000 cGMP molecules. The corresponding decrease of the cGMP concentration has the effect of reducing the flow of sodium ions into the rod. This results in an electric excitation that propagates, in modified form, along the optical nerve, passes the lateral geniculate body acting as a neuronal 'switching station', and finally arrives at the visual cortex where it is transformed, in a mysterious manner, into visual perception. Note that the choice of rhodopsin by nature was a stroke of genius in itself!

1.3 Indeterminism

Classical physics is reigned by determinism. If you know the initial conditions for a system and the forces acting upon it, you can predict its future behaviour *with certainty*. Therefore, it came as a surprise when radioactive decay processes – the emission of α particles (He nuclei), β particles (electrons) or γ quanta from nuclei – were observed to happen at random instants. Only the behaviour of a large ensemble of such nuclei follows a simple, namely exponential, decay law: the number of nuclei, $dN(t)$, decaying in a time interval from t to $t + dt$ is given by

$$dN(t) = N_0 \Gamma e^{-\Gamma t} dt, \tag{1.1}$$

where N_0 is the number of (excited) nuclei at $t = 0$ and Γ is the inverse of the mean lifetime T, $\Gamma = 1/T$. Hence we must content ourselves with a *statistical* description of an *individual* nucleus. All we can say is that it will decay with the probability

$$dw(t) = \Gamma e^{-\Gamma t} dt \tag{1.2}$$

during the time interval from t to $t + dt$.

We know for sure that a decay will take place at a certain time; we are unable, however, to predict this moment. Amazingly, though we are dealing with microscopic processes, their timescale is manifestly macroscopic. In fact, the mean lifetimes T can be as long as years, or even hundreds or thousands of years. So a few pert nuclei will decay very soon compared with T, while others are lazy and take their time over decaying, some even living (remaining in the excited state) longer than T.

A second example of indeterminism is provided by beamsplitting. This process is well known from optics: a partly transmitting mirror divides an incident beam into two beams, a reflected and a transmitted one. The same job is done by crystals for neutrons and by intense standing electromagnetic waves for atoms. Let us consider the situation when one particle after the other is falling on a beamsplitter with equal reflectivity r and transmittivity t ($r = t = \frac{1}{2}$). It is clear what will happen. Since particles are indivisible, an individual particle can only be reflected or transmitted *as a whole*, and both events will occur with 50 per cent probability. This prediction is readily confirmed by placing a detector in each output channel. They will never respond jointly, i.e., no coincidences will be observed.

It may be felt as a surprise, however, that photons – the elementary 'constituents', with energy $h\nu$ (h Planck's constant and ν frequency), of electromagnetic (especially light) fields – behave exactly in the same manner (in the aforementioned experimental conditions), thus demonstrating their particle nature. What happens at the beamsplitter is a completely random 'decision' of each photon which way to go. Needless to say that such a particle-like behaviour is in conflict with the wave picture we are accustomed to associate with optical phenomena. In fact, in classical wave theory energy spreads over the whole spatial region that is filled with the electromagnetic field. Hence, when the ingoing field contains the energy of just one photon, $h\nu$, the two partial waves emerging from the beamsplitter possess only half this energy. This is, however, not enough to trigger a photodetector – it requires the full energy of a photon. So what one would expect from classical wave theory is that *neither* photodetector responds. Even worse, such 'half photons' generated by the beamsplitter could not be absorbed by any kind of known matter, since the elementary absorption process, taking place in atoms or molecules, is governed by quantum rules. So 'half photons', or more generally any 'fractions' of photons, would say good-bye to our world, living a completely undisturbed life – indeed, a disquieting idea. Fortunately, experiment confirms the particle picture of light. So, what we have to do is to revise the classical wave concept!

The beamsplitting process discussed before can be used to generate random numbers. A signal from, say, the first detector gives us a binary digit '1' and a signal from the second detector a digit '0'. Thus we get a sequence of truly random numbers. Actually, the same cannot be guaranteed, in principle, by known random number generators. When they are macroscopic devices, they are subjected to the deterministic laws of classical physics. For instance, the apparent randomness to be observed in throwing dice results from an (assumed!) randomness of the initial conditions. Precise knowledge of the latter would enable us to predict the outcome of any throw. Even the usual procedure of generating random numbers with the help of computers is completely deterministic – it is based on suitable algorithms (e.g., decimal digits of π are calculated). Hence one cannot safely exclude the possibility that certain regularities are hidden in the calculated sequences.

Randomness means that we cannot find a cause that can be made responsible for the respective event. Such a kind of behaviour is hard to believe in, since it is in sharp contrast to classical physics and, hence, also our everyday experience. Notably Einstein could not put up with the idea of indeterminism, 'the Old One does not throw dice' being his credo. However, in Newton's words, there is no arguing against facts.

It is interesting to note that Newton found himself confronted with the unpredictable behaviour of the light particles he had postulated. What causes such a particle to be reflected, and another one to be transmitted, when hitting the surface of a transparent medium? To give an answer, Newton resorted to the idea that particles are in different fits: a particle being in a 'fit of easy reflection' will become reflected, while a particle in a 'fit of easy transmission' will be transmitted. With this concept, Newton was actually the first to postulate the existence of what was later called hidden variables!

Anyway, it is disquieting to see that microscopic randomness actually intrudes, through measuring apparatuses, in the macroscopic world. In fact, a signal from any detector can be utilized to initiate a second process, thus starting a causal chain of macroscopic events. Hence, already from this reason the idea of a completely deterministic world cannot be upheld. Let us consider, for illustration, an experimental set-up that would work as a truly probabilistic 'Russian roulette'. Actually, it is an up-to-date version of Schrödinger's 'diabolic device' destined to kill, fortunately only in thought, an innocent cat (for details see Section 5.1).

The idea is to use the radioactive decay of a nucleus to trigger, via its detection, the explosion of a bomb. With the half-life period of the decay assumed to be approximately an hour, a person sitting close to the bomb, say for half an hour, might get satisfaction from the idea that inconceivable fate itself will decide on his 'being or not being'.

1.4 Wave–particle dualism and a new kind of uncertainty

Interference effects rank among the most fascinating optical phenomena. Even nature utilizes interference techniques. For instance, butterflies owe their wonderful colours to interference. From the physical point of view, interference results from the superposition of two (or more) fields. Formally, superposition is a consequence of the linearity of the wave equation: if $A(r, t)$ and $B(r, t)$ are two (in general complex) solutions of the wave equation, then the sum $C(r, t) = \alpha A(r, t) + \beta B(r, t)$, where α and β are arbitrary complex numbers, is a solution too.

In Young's pioneering interference experiment, a screen with two pinholes (or slits) is illuminated with a point-like light source, and the light coming from one pinhole gets superposed with the light from the other (see Fig. 1.1). As a result, interference fringes show up on an observation screen. In an interferometer, on the other hand, an incident light beam is divided, with the help of a beamsplitter, into two coherent beams of equal intensity, 1 and 2. They travel along different interferometer arms and are eventually reunited at the output mirror. Consider for instance a Mach–Zehnder interferometer (see Fig. 1.2). At the output mirror the reflected part of beam 1 interferes with the transmitted part of beam 2, and similarly the transmitted part of beam 1 interferes with the reflected part of beam 2. Variation of one of the arm lengths leads to a change of the intensities of the two outgoing beams. To be more specific, the ratio of those intensities varies periodically as $\cos^2(\Delta s/\lambda)$, where Δs is the geometrical difference in the arm lengths and λ is the wavelength of the light. In particular, the interferometer can be adjusted such that all the incident energy leaves it in just one output channel.

This is all well known from classical optics. But what will happen when the incident beam is attenuated more and more? It follows from the linearity of Maxwell's equations that the interference pattern will persist without any degradation, however small the intensity might be made. (To compensate for the intensity loss, the exposure time must be enhanced accordingly.) This was confirmed experimentally as early as 1909 by Taylor (1909), who made observations with exposure times up to 3 months. Nevertheless, this result becomes puzzling in the photon picture. Very low intensity means that photons arrive, one after the other, with a wide separation. Actually, the average temporal distance between subsequent photons can readily be made much larger than the passage time through

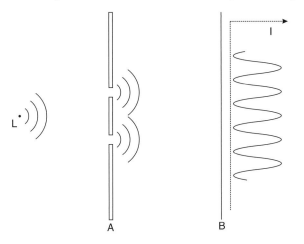

Fig. 1.1 Young's interference experiment. L = light source; A = interference screen; B = observation screen; I = intensity distribution.

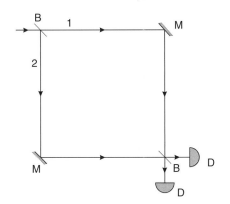

Fig. 1.2 Mach–Zehnder interferometer. B = beamsplitter; M = deviating mirror; D = detector.

the interferometric device. Then we can be sure that at any time there is one photon, at most, present in the apparatus. Thus we are led to the conclusion that every photon interferes with itself, as Dirac (1958) formulated it.

It should be noticed, however, that interference is actually a collective phenomenon. In fact, a single photon produces, at best, a single blackened spot on a photographic film, and we cannot say whether this spot will become a constituent of an interference pattern or of a more or less homogeneously blackened area. The situation is similar in a Mach–Zehnder interferometer, for instance. Even when it is adjusted such that only one output channel is open, the detection of a single photon in this channel does not prove that the other must be blocked. In fact, we cannot exclude the possibility that the observed event happened by mere chance, the interferometer being actually misadjusted. To check the adjustment, we must send many photons through the interferometer. When the adjustment is correct, *all* photons are bound to exit in the allowed channel. On the other hand, it suffices that only one photon appears in the forbidden channel to indicate misadjustment. So we need to observe many photons to be able to detect interference. Interestingly, this is not so in experiments demonstrating the corpuscular character of particles, including the photon, where just one particle gives us full evidence (see Section 1.3). Although interference can be observed only on an ensemble of photons, every individual photon must, of course, behave in accordance with the interference pattern that eventually evolves. In particular, it must strictly obey the prohibition of going to the zeros of intensity on an interference screen or exiting through a forbidden output channel.

Amazingly, interference experiments can be (and actually were) performed also with massive particles such as electrons, neutrons and even atoms or molecules. The outcomes are quite similar to what is observed with photons (also here, each particle interferes with itself), the wavelength being now given, after de Broglie, by $\Lambda = h/p$, where p is the particle's momentum. In particular, a Mach–Zehnder interferometer for neutrons can be realized. Hence we cannot but acknowledge the fact that elementary microscopic objects, such as photons on the one hand and massive particles on the other hand, behave like classical particles *or* classical waves, depending on the experimental conditions. This striking feature is known as wave–particle dualism. So what we call elementary particles are not particles in the classical sense; we should rather say they are nonclassical objects that have *both* corpuscular and wave-like properties. In particular, in an interference experiment both aspects, the wave-like and the particle-like, become manifest. The wave-like aspect shows up in the interference pattern, whereas the detection process bears witness to the particle-like aspect. In fact, an interference pattern results from individual events in which a particle (or a photon) interacts locally, e.g., with a photographic plate.

However, the interference of a particle with itself is hard to understand. Prejudiced by our everyday experience, we will argue: a particle can pass only one pinhole in Young's interference experiment or go only one way in an interferometer. How can it then know that there is a second pinhole or a second interferometer arm? Quantum theory shows a way out of this dilemma: we must give up the idea that a particle always travels along a definite path. Instead, we must ascribe to it an intrinsic position uncertainty which forbids us to think of it as going one way or the other. Strikingly, this uncertainty is not necessarily restricted to microscopic dimensions (which would not disturb us too much); it may, in fact, spread over manifestly macroscopic distances. This is so in interferometers, the record being presently held by Michelson interferometers destined to detect gravitational radiation, whose arm lengths attain hundreds of metres. This intrinsically quantum mechanical uncertainty may be considered as a counterpart of what is known as mutual coherence (between different beams) in classical optics. From the viewpoint of quantum mechanics, the ability to interfere goes hand in hand with uncertainty of the particle's path. On the other hand, when the way the particle has taken is known, no interference takes place. Actually, it is not necessary that we get the 'which way' information from a measurement, it suffices that the particles are subjected to an interaction that *might* serve as the heart of a possible measuring scheme. Really striking are interference experiments with atoms, where internal degrees of freedom are used to encode 'which way' information (Dürr, Nonn and Rempe, 1998).

It should be emphasized that the quantum mechanical uncertainty in question is a new physical quality that has no counterpart in classical physics. By no means should it be confounded with the classical uncertainty concept. A basic credo of a classical physicist is that any physical variable has a perfectly sharp value, at any time, irrespective of whether we have measured (or even are able to measure) it or not. Obviously, nobody can ever verify experimentally this assumption in full generality; the point is, however, that in classical physics we can afford such a philosophy, since it is compatible with all our experience. With this concept of 'objective reality', uncertainty becomes a subjective category: it results only from our inability to procure the necessary information. In other words, uncertainty is nothing but a lack of knowledge, and this is in sharp contrast to what is understood by uncertainty in quantum theory.

1.5 Quantized energies

Spectroscopists provided us with exciting news of the world of atoms. They found out that such tiny objects, when excited, emit light only at discrete frequencies, the so-called spectral lines, characteristic of their respective species. This is, in fact, puzzling and, still worse, incompatible with atomic models based on classical physics. It was Bohr who detected the key to solve this riddle, putting forward the revolutionary idea that only selected electronic orbits might be allowed as stationary atomic states. He conceived the processes of light emission and absorption as being connected with abrupt transitions between two stationary atomic states with different energies E_m and E_n, respectively. So the idea of 'quantum jumps' was born. Bohr postulated a resonance condition for the frequency ν_{mn} of the emitted, or absorbed, light in the form

$$E_m - E_n = \pm h\nu_{mn}, \tag{1.3}$$

which establishes a fundamental connection between atomic properties and emission as well as absorption spectra. When E_m is the energy of the initial state and E_n that of the final state, the plus sign holds for emission ($E_m - E_n > 0$) and the minus sign for absorption ($E_m - E_n < 0$). Thus, with the help of spectroscopic measurements we get a wealth of information on atoms in the form of energy level structures, the so-called term schemes. This makes it evident that the quantization concept put forward first by Planck, albeit reluctantly, and independently inferred by Einstein from a thermodynamic analysis of the radiation field, proves to be an invaluable clue to the mysteries of the microscopic world.

2

Quantum states

In this chapter the quantum mechanical state concept will be discussed in some detail. First, I would like to remind the reader of the mathematical principles on which quantum theory is based.

2.1 Elements of the quantum mechanical formalism

The novel kind of uncertainty we encountered in Section 1.4 requires a new mathematical apparatus for its proper description. Heisenberg had the ingenious idea of associating position x and momentum p of a particle[1] with mathematical objects that do not commute. These can be identified with Hermitian matrices or, more generally, linear Hermitian operators $\hat{x}$ and $\hat{p}$. Postulating the commutation relation

$$\left[\hat{x}, \hat{p}\right] \equiv \hat{x}\hat{p} - \hat{p}\hat{x} = i\hbar\mathbf{1} \tag{2.1}$$

($\hbar = h/2\pi$), he laid the foundations of modern quantum theory. The appearance of $\hbar$ in the commutation relation indicates the fundamental role Planck's constant plays in quantum physics.

With the variables x and p all (classical) physical quantities that are functions of x and p become operators too. This applies, in particular, to the energy, for instance the energy (sum of kinetic and potential energy) of an electron moving in the Coulomb field of a proton (hydrogen atom). It appears natural to adopt quantization as a general rule: in the quantum mechanical description, any measurable quantity A, a so-called observable, is represented by a linear Hermitian operator $\hat{A}$. It is important to note that explicit expressions for energies, in particular interaction energies, are taken from classical physics. Hence quantum theory is founded on the experience we have gathered in the *macroscopic* world.

---

[1] The quantities x and p are to be understood as Cartesian components, with respect to the same direction, of the position and the momentum vectors.

Operators need objects to act on. In Schrödinger's theory, these are integrable complex space functions $\psi(x)$ $(x = (x_1, x_2, x_3))$, and the operator $\hat{x}_j$ means multiplication by x_j $(j = 1, 2, 3)$, while the operator $\hat{p}_j$ is given by the differential operator $-i\hbar\partial/\partial x_j$. When the temporal evolution is taken into account, the Schrödinger functions become time-dependent. They obey a new form of a wave equation (Schrödinger equation) and hence apparently describe some kind of matter waves. It must be emphasized, however, that, unlike the fields we know from classical physics (for instance, electromagnetic waves), they do not directly describe physical reality. This becomes obvious when one attempts to interpret (as Schrödinger really did in his first papers) the quantity $e\,|\psi|^2$ (e elementary electric charge) as the charge density for an electron. So the electron would not be a point-like particle, but rather 'smeared out' someway. However, with this concept we run into serious difficulties. In fact, initially localized wavefunctions normally spread more and more during their evolution. A way out of this dilemma was shown by M. Born, who suggested a statistical interpretation of the wavefunction: $|\psi(x)|^2 d\tau$ is the *probability* of finding the electron, when measured, in a volume element $d\tau$ centred at position x. So the wavefunction is a rather abstract entity. Nevertheless, Schrödinger's concept is too special, since it favours the position variable. Indeed, later on a general quantum mechanical formalism was developed (von Neumann, 1932) that gave quantum mechanics a sound mathematical basis. It rests on the concept of an abstract space named Hilbert space.

A Hilbert space $\mathcal{H}$ is a linear space: When $|\varphi\rangle$ and $|\psi\rangle$ are elements (vectors) of $\mathcal{H}$, then also the linear combination $|\chi\rangle = \alpha\,|\varphi\rangle + \beta\,|\psi\rangle$ (α, β complex numbers) belongs to $\mathcal{H}$. An important property of a Hilbert space is that a (complex) scalar product $\langle\varphi|\,\psi\rangle\,(= \langle\psi|\,\varphi\rangle^*)$ is defined for any two vectors $|\varphi\rangle$ and $|\psi\rangle$. Specializing to $|\psi\rangle = |\varphi\rangle$, the scalar product $\langle\varphi|\,\varphi\rangle$ gives us the square of the norm of the vector $|\varphi\rangle$. With the scalar product at hand, we are able to define an orthogonality relation: two vectors are orthogonal, when their scalar product vanishes. Moreover, any vector can be normalized to unity, $\langle\varphi|\,\varphi\rangle = 1$. Normalized vectors are used in quantum theory to characterize the physical states in which a system might be. In this function, they are called state vectors.

In a Hilbert space, linear operators $\hat{B}$ can be defined that transform a vector $|\varphi\rangle$ into a vector $|\varphi'\rangle = \hat{B}\,|\varphi\rangle$. Of special physical interest are vectors that are reproduced, apart from a factor, under the action of a chosen Hermitian operator $\hat{A}$:

$$\hat{A}\,|a_j\rangle = a_j\,|a_j\rangle. \tag{2.2}$$

The vectors $|a_j\rangle$ were named eigenvectors of $\hat{A}$, and the as are the corresponding eigenvalues.

It follows from the Hermiticity of $\hat{A}$ that (i) the eigenvalues are real and (ii) that eigenvectors corresponding to different eigenvalues are orthogonal. When two or more linearly independent eigenvectors correspond to the same eigenvalue – we then speak of a degenerate eigenvalue – they can always be chosen as mutually orthogonal. Hence the eigenvectors altogether form an orthogonal system. The latter is complete; this means that any vector $|\psi\rangle$ can be expanded in terms of eigenvectors, which we assume to be normalized

$$|\psi\rangle = \sum_j c_j |a_j\rangle, \quad c_j = \langle a_j| \psi\rangle. \tag{2.3}$$

(The expression for c_j follows directly from the orthonormality of the system of eigenvectors.)

The eigenvalues and eigenvectors have a direct physical meaning: as was mentioned above, any measurable physical quantity (observable) A is represented by a Hermitian operator $\hat{A}$. Its eigenvalues are the possible outcomes of a measurement of A, and the squared modulus of the expansion coefficient, $|c_j|^2 = |\langle a_j|\psi\rangle|^2$, gives us the probability that the measurement result is a_j, when the system is initially in the state $|\psi\rangle$. It should be noted that not all observables have discrete spectra. Notably, the operators for position and momentum possess a continuous spectrum each, extending from $-\infty$ to $+\infty$. Then the rule predicting the outcome of a measurement has the above-mentioned form introduced by Born.

Only when the system is already in an eigenstate $|\varphi_k\rangle$ of $\hat{A}$ can the outcome of a measurement of the observable A be predicted with certainty. (We then have $|c_k|^2 = 1$ and $|c_j|^2 = 0$ for $j \neq k$.) In general, the indeterministic character of the quantum world shows up: it is by chance what comes out in a measurement on an individual system.

Of special physical interest is the average over the measured values, the so-called expectation value of the observable A,

$$\langle A\rangle = \sum_j |c_j|^2 a_j, \tag{2.4}$$

which can be rewritten, thanks to the orthogonality of the eigenvectors, as

$$\langle A\rangle = \langle \psi| \hat{A} |\psi\rangle, \tag{2.5}$$

where the right-hand side of this equation is to be understood as the scalar product of the vectors $|\psi\rangle$ and $\hat{A} |\psi\rangle$ or, equivalently, of $\hat{A} |\psi\rangle$ (since $\hat{A}$ is Hermitian) and $|\psi\rangle$.

As is known from classical statistics, an appropriate measure of the uncertainty of a variable is the mean square root deviation. Accordingly, we define the quantum mechanical uncertainty of an observable A, ΔA, by

$$(\Delta A)^2 = \left\langle A^2 \right\rangle - \langle A \rangle^2 = \langle \psi | \hat{A}^2 | \psi \rangle - \langle \psi | \hat{A} | \psi \rangle^2. \tag{2.6}$$

We know from classical physics that different variables can be measured simultaneously on a single system. In quantum theory, on the contrary, joint measurements frequently cause trouble. The formal reason is that Hermitian operators have common eigenstates then, and only then, when they commute. In this case the corresponding observables can, in fact, be measured simultaneously on one and the same system. However, this is not possible when the operators do not commute. This applies, in particular, to canonically conjugate variables, since they are subjected to canonical commutation rules of the type given in (2.1).

In the case of noncommuting operators, one can nevertheless measure the corresponding observables simultaneously, however, not on the same system but on different subensembles. For instance, you measure position x on one subensemble that is arbitrarily selected from the given total ensemble, and momentum p on another. From the measured data you can easily determine the uncertainties Δx and Δp, respectively. Now, it is of great physical relevance that those quantities fulfil what has been called an uncertainty relation. In fact, it can rigorously be shown that the commutation relation (2.1) implies the inequality

$$\Delta x \Delta p \geq \hbar/2, \tag{2.7}$$

which became famous as Heisenberg's uncertainty relation (cf. Pauli, 1933).

It should be mentioned that simultaneous measurements (on the same system) might actually be carried out with the help of 'meters'. However, this can be achieved only at the cost of measurement precision, as will be explained in some detail in Section 3.5.

Finally, let me say a few words about the time evolution of a quantum system. In the Schrödinger picture the wavefunction, or, more generally, the state vector is assumed to be time-dependent. It obeys a first-order linear differential equation, the famous Schrödinger equation

$$i\hbar \frac{\partial |\psi\rangle}{\partial t} = \hat{H} |\psi\rangle, \tag{2.8}$$

where $\hat{H}$ is the Hamilton operator (Hamiltonian) of the system, and is obtained from the classical Hamilton function by replacing the conjugate variables by Hermitian operators. Since the Schrödinger equation is of first order with respect to t, the initial condition is rather simple: it requires the solution to match the initial state vector. Though quantum mechanics is intrinsically indeterministic, the evolution of the state vector is a completely deterministic process!

Equation (2.8) describes not only the free motion (the system is left to itself), it applies also when the system is exposed to external forces or fields, or when two or more systems interact. In the first case the Hamiltonian becomes explicitly time-dependent, and in the latter case it splits into two parts, an undisturbed Hamiltonian $\hat{H}_0$ governing the free motion of the individual systems, and an interaction Hamiltonian $\hat{H}_{int}$ that accounts for their coupling. In the interaction case, the Schrödinger equation can usually be solved only approximately, mostly with the help of perturbation theory.

Formally, it is easy to write down the solution in the energy representation, i.e., by expanding the state vector in terms of energy eigenstates

$$|\psi\rangle = \sum_k c_k \, |E_k\rangle \, . \tag{2.9}$$

Actually, the time dependence of an energy eigenvector $|E_k\rangle$ follows from (2.8) to be given by

$$|E_k\rangle_t = e^{-i\omega_k(t-t_0)} \, |E_k\rangle_{t_0} \, , \qquad \omega_k = \frac{E_k}{\hbar}, \tag{2.10}$$

and thanks to the linearity of the Schrödinger equation we have only to replace $|E_k\rangle$ by $|E_k\rangle_t$ in (2.9) to get the (exact!) solution that satisfies the initial condition. The problem is, however, only reduced to the task of diagonalizing the (total) Hamiltonian (i.e., finding its eigenvalues and eigenvectors), which is anything but simple in the case of interacting systems.

It should be noted that the quantum mechanical time evolution is a unitary transformation

$$|\psi\rangle_t = \hat{U}(t, t_0) \, |\psi\rangle_{t_0} \, . \tag{2.11}$$

Here, $\hat{U}$ is a unitary operator, i.e., together with its Hermitian adjugate $\hat{U}^\dagger$ it satisfies the relations $\hat{U}\hat{U}^\dagger = \hat{U}^\dagger\hat{U} = \mathbf{1}$. It follows from the equation of motion (2.8) that $\hat{U}$ fulfils the differential equation

$$i\hbar \frac{\partial}{\partial t} \hat{U}(t, t_0) = \hat{H}(t)\hat{U}(t, t_0) \tag{2.12}$$

with the initial condition $\hat{U}(t_0, t_0) = \mathbf{1}$. A solution for the evolution operator can be found in the form

$$\hat{U}(t, t_0) = \mathbf{1} + \frac{1}{i\hbar} \int_{t_0}^{t} dt_1 \hat{H}(t_1) + \cdots + \frac{1}{(i\hbar)^n} \int_{t_0}^{t} dt_1 \ldots \int_{t_0}^{t_{n-1}} dt_n \hat{H}(t_1) \ldots \hat{H}(t_n) + \ldots,$$

$$\tag{2.13}$$

where we have assumed that the Hamiltonian is explicitly time-dependent. In the case of conservative systems the Hamiltonian does not explicitly depend on time. Then the time integrations in (2.13) are easily done to yield the simpler expression

$$\hat{U}(t,t_0) = \mathbf{1} + \frac{\hat{H}}{i\hbar}(t-t_0) + \cdots + \frac{\hat{H}^n}{(i\hbar)^n}\frac{(t-t_0)^n}{n!} + \cdots = \exp[-\frac{i}{\hbar}\hat{H}(t-t_0)].$$

$$(2.14)$$

An alternative, but completely equivalent way of describing the evolution, known as the Heisenberg picture, is to consider the observables as time-dependent, whereas the state vector characterizes the initial conditions and, hence, does not depend on time. The equation of motion for an observable A reads

$$i\hbar\frac{\partial\hat{A}}{\partial t} = \left[\hat{A}, \hat{H}\right].$$

$$(2.15)$$

Note that Heisenberg's equations of motion nicely correspond to the classical description: they are obtained from the classical equations of motion by replacing the Poisson brackets by the corresponding commutators, multiplied by $-i/\hbar$. Actually, the Heisenberg picture has the great advantage that it allows us to calculate two-time correlation functions in a simple way.

Whereas the evolution considered thus far is a continuous process, any measurement gives rise to an abrupt change of the state vector: it reduces to that part of its expansion (2.3) in terms of eigenvectors of the observable in question that is compatible with the outcome of the measurement. This is indeed a necessary condition for an ideal measurement, since we have to require that an immediate repetition of the measurement yields the same result as before, with certainty. The probability of finding just the eigenvalue a_j as a measurement result is given by $w_j = |c_j|^2$, but nobody can tell which value will actually be measured in a single case. Quantum mechanics assures that, apart from the constraints set by the probabilities w_j, the measuring process is governed by mere chance. We defer a more thorough discussion of the reduction rule to Section 3.6.3.

From what has been said above, it becomes clear that quantum mechanics focuses on what happens when microscopic systems are subjected to observations carried out with macroscopic measuring apparatuses. In view of the complexity of such measuring processes the success of quantum mechanics in describing (in particular, predicting) any kind of experiment borders on the miraculous. One should be aware, however, that the price for this achievement is a radical break with the philosophy that underlies classical physics. While the latter deals with physical entities (particles and objects formed from them, on the one hand, and forces, on the other hand), quantum theory is based on an entirely new concept: its formalism describes, in the first place, potentiality rather than reality. In other words, primarily it has to do with the possible rather than the factual. However, the proper instrument to handle quantum mechanical potentiality is not simply probabilities but probability *amplitudes*, as they appear in (2.3). Since they are complex numbers, they contain – through their phases – much more information than the probabilities (which are the squared moduli of the amplitudes). This feature is, in fact, crucial to the quantum mechanical formalism.

In particular, Schrödinger's wavefunction has the meaning of a (space and time dependent) probability amplitude for position measurements. It accounts for the wave-like behaviour of massive particles. Interference phenomena result now from the superposition of probability amplitudes, in close analogy to (classical) optical interference. However, the classical wavefunction describes a quantity that is measurable (at least in principle), namely the electric field strength, whereas the quantum mechanical wavefunction has the meaning of a probability amplitude. So we have the following situation. In classical optics, we understand the physical mechanism of interference, e.g., in an interferometer electric field strengths residing at the same position will add, thus enlarging or diminishing the total electric field strength (in dependence on the relative phase), as we know it quite generally from any kind of forces. (At school you learned from tug of war that two forces might compensate.) The difference of the arm lengths comes into play, because two ('really existing') partial waves propagate, each along one of the arms. In particle interference, on the contrary, the part of the electric field strength is played by something mysterious – the probability amplitude – and we are left alone with the question as to which *physical* mechanism underlies the interference phenomenon, and, in particular, how the particle manages to get the necessary information about the difference of the arm lengths. As mentioned in Section 1.4, we must content ourselves with the statement that quantum mechanical uncertainty indicated by the transverse spatial extension of the wavefunction (in the interferometer case it covers both arms) is a prerequisite for interference to be observed.

2.2 Pure states

Let us consider a simple quantum mechanical experiment that shows up the intrinsically indeterministic character of the microscopic world. We produce single polarized photons by strongly attenuating high-intensity (classical) quasimonochromatic light that has been linearly polarized, say in the x direction, with the help of a polarizing prism. The photons follow one after the other at irregularly distributed distances assumed large enough to allow for individual observations. A quantum theorist would say, we have 'prepared' photons in the x polarization state.

Now we send the photons through a second polarizing prism with a (unit-efficiency) detector placed in each exit. This prism shall be rotated, however, by an angle Θ with respect to the orientation of the first polarizer, its directions of transmittance being now x' and y'. We can readily say what will happen. Since the photon's energy $h\nu$ cannot be divided into smaller parts, at any time just one detector can respond, but never both. Making a lot of individual measurements, say N, we will observe N_1 clicks from detector 1 indicating x' polarization and $N_2 (= N - N_1)$ clicks from detector 2 indicating y' polarization. From those data we infer the probabilities of measuring x' or y' polarization to be $w_1 = N_1/N$ and $w_2 = N_2/N$. Since a large number of photons will behave like a classical field, we can take the values of the probabilities from classical optics: a polarizing prism projects the electric field strength vector on the x' and the y' direction, respectively. Hence the intensities of the outgoing beams are given by $I_1 = \cos^2 \Theta I$ and $I_2 = \sin^2 \Theta I$, respectively, where I is the intensity of the ingoing beam. Now, the intensity is nothing but the number of incident photons per second, multiplied by $h\nu$, and so we can specify the quantum mechanical probabilities as $w_1 = \cos^2 \Theta$ and $w_2 = \sin^2 \Theta$. It is interesting to note that our simple device allows us, in fact, to perform a variety of measurements; we need only change the angle Θ.

A classical physicist would see the reason for the different behaviour of the incoming photons in a difference of their physical states. He would postulate the existence of a physical parameter that can take different values, thus determining whether an individual photon takes one way or the other in the polarizing prism. However, he won't be able to specify such a variable. To save his deterministic philosophy he would be forced to resort to *hidden* variables that cannot be measured, on principle. Quantum theorists, however, being fully convinced that the microscopic world is governed by pure chance, rejected the idea of hidden variables. Instead they developed the following state concept: when a large number of systems were prepared in the same way – this means, they have emerged from a macroscopic device whose parameters were precisely set – they form an ensemble that is in a pure quantum state. Such an ensemble is homogeneous, i.e., it is impossible to select subensembles that can be distinguished physically from one another and the original ensemble. In particular, measurements of *any* observable, when carried out on different subensembles, will yield identical frequency distributions for the possible outcomes. To come back to our example, it is in particular impossible to split the ensemble of photons beforehand in two subensembles such that all members of the first ensemble will be detected as x' polarized, whereas those belonging to the second ensemble will be detected as y' polarized.

Since a quantum state describes basically a whole ensemble of equally prepared systems, we are left with the question how to deal theoretically with a single system. I will postpone the discussion of this problem to Section 2.5.

As was already mentioned in Section 2.1, a pure state is described by a Schrödinger wavefunction or, more generally, a (normalized) vector in Hilbert space $|\psi\rangle$. It looks like magic to see what a wealth of information is wrapped in it. Following the instructions given in Section 2.1, we are able to literally predict all that might be observed. (*i*) Given *any* observable whose eigenvalues a_k and eigenvectors $|a_k\rangle$ we know, we get the probability of just finding the value a_k as the outcome of the measurement by forming the squared modulus of the scalar product $\langle a_k | \psi\rangle$. (*ii*) We are able to calculate expectation values of arbitrary Hermitian operators (see (2.5)), among them arbitrary powers of a given operator. In particular, quantum mechanical uncertainties, as defined in (2.6), can thus be evaluated. (*iii*) It is also possible to calculate from the state vector correlation functions of any kind, for instance, electric field correlations that determine the interference behaviour. A second example is intensity correlations, i.e., coincidence counting rates of two photodetectors placed at different positions in a radiation field. (Those events are registered when the detectors click simultaneously, or, in the case of delayed coincidences, when the second detector responds a given time later than the first.) (*iv*) Knowing the state vector at a certain time, we can also study its time evolution (see (2.8) and (2.15)). Since the state vector at a later time is *uniquely* determined by the equation of motion, ensembles evolve deterministically. Thus ensembles that are identically prepared and subjected to the same external influences will also remain identical in the future. So, as in classical physics, the future can be reliably predicted. This applies, however, only to ensembles and not to individual systems! Of special interest is the case of interacting systems. Then we get full information on what will happen, thanks to the interaction.

It is important to note that the quantum mechanical state concept enables us to give the intrinsically quantum mechanical uncertainty explained in Section 1.4 a precise mathematical form. We just have to form superposition states! For instance, the superposition

$$|\psi\rangle = c_1 |E_1\rangle + c_2 |E_2\rangle , \qquad (2.16)$$

where $|E_1\rangle$ and $|E_2\rangle$ symbolize the ground state and the first excited state, respectively, of an atom, describes an atomic state with undefined energy. Such a state can be realized experimentally by shining a resonant light pulse on a gas of atoms being initially all in their ground states. (For more details see Section 3.3.3.)

It becomes obvious from (2.16) that the energy uncertainty is quite different from simply not knowing whether the atoms are excited or not. In fact, this uncertainty has an important physical consequence. Calculating the expectation value of the electric dipole operator $e\boldsymbol{r}$ (e electric charge of the electron and $\boldsymbol{r}$ vector pointing from the atomic centre to the electron) for the superposition state, we will get a nonvanishing term in favourable cases. Taking into account the free atomic motion, this dipole moment oscillates with the frequency given by the difference of the atomic frequencies E_2/h and E_1/h. The amplitude of the dipole oscillation becomes maximum when the moduli of the coefficients c_1 and c_2 are equal. Experimentally, this can be achieved by properly choosing the intensity of the exciting optical pulse, for a given pulse duration (see Section 3.3.3).

It is interesting to note that the superpositon (2.16) contains a phase δ, namely the relative phase between the complex numbers c_1 and c_2. This formal phase is, in fact, of physical relevance. It traces back to the phase of the exciting pulse, and it determines the phase of the induced dipole oscillation. Now, the individual dipole moments sum up to a *macroscopic* dipole moment which, in turn, acts as a source of radiation. It emits a light pulse whose phase is determined by the dipole phase. Hence the phase δ can actually be measured. All this is well known from classical physics. In quantum theory, however, energy and dipole moment are complementary variables: sharp energy and the presence of an oscillating dipole moment are mutually exclusive.

While in the analysis of atomic and molecular processes the pure quantum states of interest are mostly eigenfunctions of energy or spin, or both, the situation is different in quantum optics. The energy eigenfunctions of a single-mode radiation field are states with sharp photon numbers n, the so-called Fock states $|n\rangle$ $(n = 0, 1, 2, \ldots)$. They form a basis in Hilbert space that is very convenient for calculations; however, they suffer from the disadvantage that only Fock states with no more than two photons can be produced in experiments. Of greater physical relevance are the so-called coherent states, often named Glauber states (Glauber, 1965). They are the quantum mechanical analogues of classical waves with fixed amplitudes and phases. Actually, those variables cannot be simultaneously sharp in quantum theory, as follows from the noncommutability of the photon number operator and the electric field strength operator.

In a Glauber state, however, the fluctuations of the amplitude and the phase are 'optimized' in the sense that the uncertainty of the electric field strength becomes minimum, under the constraint that the average photon number is fixed. Expanded in terms of Fock states, the coherent states take the form

$$|\alpha\rangle = e^{-\frac{|\alpha|^2}{2}} \sum_{n=0}^{\infty} \frac{\alpha^n}{\sqrt{n!}} |n\rangle, \tag{2.17}$$

where the complex number α corresponds to the classical complex amplitude. The normalization is such that $|\alpha|^2$ equals the average photon number. As was first recognized by Glauber, coherent states can be made the basis of a quantum mechanical description that closely parallels that of classical optics. Moreover, they can easily be generated: light from a single-mode laser is in such a state, to a very good approximation, and fortunately linear damping fully preserves the property of being in a coherent state. Only the modulus of α is reduced, as it must be. As far as photon statistics is concerned, this is no matter of surprise. In fact, according to (2.17) the photons follow a Poisson distribution, and it is well known in classical statistics that such a distribution remains Poissonian when elements are taken away in a random manner. On the other hand, we know from classical optics that the phase of a light wave remains sharp when the wave is damped.

Since damping can be made as strong as one likes, starting from laser fields, coherent states can be produced in the microscopic domain. Their average photon number can, in fact, be set at will. It may be chosen, for instance, of the order of unity, but there is no lower limit: it may also take values much smaller than unity.

So coherent states are an invaluable tool for experimentalists. Their practical usability, however, goes hand in hand with mathematical peculiarities that trouble the theorists. The point is that the coherent states are *not* eigenfunctions of any Hermitian operator (which, in particular, means that α is not an eigenvalue). Hence, they do not form an orthogonal basis.

2.3 Mixed states

The idea of a mixed state is basic to classical statistics. Since we usually do not know precisely the values of all variables of a physical system – especially when the system is formed of an enormous number of constituents, as in the case of a gas, for instance – we have to content ourselves with a reduced information that is contained in distribution functions. They indicate the probability that the system is actually in a certain 'pure state', i.e., a state with given sharp values of the variables. This concept is readily extended to the quantum mechanical description. One has to put together different 'elementary' ensembles in pure states to form a more complex ensemble, which is said to be in a mixed state or a statistical mixture. Accordingly, such a more general quantum state is characterized by the weights w_k with which pure states $\left|\psi^{(k)}\right\rangle$ contribute. (The weights are just the relative numbers of systems that constitute the 'elementary' ensembles.)

We find the formal description of a mixed ensemble from considering expectation values. Let us first expand the states $\left|\psi^{(k)}\right\rangle$ with respect to an arbitrary (orthogonal) basis in Hilbert space, $|\varphi_i\rangle$,

$$\left|\psi^{(k)}\right\rangle = \sum_i c_i^{(k)} |\varphi_i\rangle . \tag{2.18}$$

Then the expectation value of any observable $\hat{A}$, for the state $\left| \psi^{(k)} \right\rangle$, can be written in the form

$$\left\langle \psi^{(k)} \left| \hat{A} \right| \psi^{(k)} \right\rangle = \sum_{i,j} \langle i | \hat{A} | j \rangle \, c_j^{(k)} c_i^{(k)*}, \tag{2.19}$$

where $\langle j | \hat{A} | i \rangle$ stands for $\left\langle \varphi_j \left| \hat{A} \right| \varphi_i \right\rangle$. To obtain the expectation value for the whole ensemble, we simply have to average (2.19) over the states $\left| \psi^{(k)} \right\rangle$. The result can be written as

$$\left\langle \hat{A} \right\rangle_{\text{mixture}} = \sum_{i,j} \langle i | \hat{A} | j \rangle \, \langle j | \hat{\rho} | i \rangle, \tag{2.20}$$

where

$$\langle j | \hat{\rho} | i \rangle = \sum_{k} w_k c_j^{(k)} c_i^{(k)*}. \tag{2.21}$$

The quantities $\langle j | \hat{\rho} | i \rangle$ can be identified with the matrix elements, with respect to the basis $| \varphi_i \rangle$, of a Hermitian operator $\hat{\rho}$ that was named density operator.

The double sum in (2.20) is known as trace operation. So (2.20) can be written in the compact form

$$\left\langle \hat{A} \right\rangle_{\text{mixture}} = \text{Tr}(\hat{A} \hat{\rho}). \tag{2.22}$$

So we have learned that the proper mathematical tool to describe a mixed state, which is, in fact, the most general quantum state, is a density operator. Clearly, in mixtures both the intrinsic quantum mechanical uncertainty (inherent in pure states) and classical uncertainty (in the sense of a lack of knowledge) are involved.

Since we know how 'elementary' ensembles evolve in time, we know also the evolution of a mixed ensemble. From the Schrödinger equation (2.8) follows the equation of motion for the density operator as

$$i\hbar \frac{\partial \hat{\rho}}{\partial t} = \left[\hat{H}, \hat{\rho} \right]. \tag{2.23}$$

Mixtures emerge quite naturally from (ideal) measurements. Indeed, when measuring any observable on an ensemble that is in a pure state, this ensemble will become split into subensembles that are in eigenstates of the observable. When we do not separate those ensembles, or simply take no notice of the outcomes of the measurement, we have an ensemble that is in a mixed state. In some cases we may actually be able to separate spatially the subensembles in the experiment, thus 'decomposing' the mixed state into pure states. Normally, the measuring apparatus does this separation by itself. Think of the polarizing prism mentioned in the beginning of this section or a Stern–Gerlach apparatus that splits an incoming beam of spin $\frac{1}{2}$ atoms into two diverging beams corresponding to the spin components $\frac{1}{2}$ and $-\frac{1}{2}$, respectively. (For more details see Section 3.1.1.)

On the other hand, any mixture can always be decomposed in thought. Amazingly, such a decomposition is not unique. This becomes obvious from the following examples.

(*i*) We consider an ensemble of photons that behaves like unpolarized light. With respect to a basis given by two orthogonal polarization states $|x\rangle$ and $|y\rangle$, the density matrix is just the (two-dimensional) unity matrix, multiplied by a normalization factor $\frac{1}{2}$. This form of the density matrix certainly allows the interpretation that the whole ensemble consists of two subensembles (with equal numbers of members) that are in the states $|x\rangle$ and $|y\rangle$, respectively. However, we may equally well choose a basis corresponding to polarization directions that are rotated, with respect to the x, y directions, by an arbitrary angle. Moreover, we may even choose any two orthogonal *elliptic* polarization states, in particular the states of left- and right-handed circular polarization, as a basis, which suggests a decomposition of the density operator into two states characteristic of elliptic (in particular circular) polarization. So there actually exists a virtually infinite number of different decompositions. In fact, all those decompositions can also be done experimentally with the help of appropriate analyzers (polarizing prisms equipped with optical elements known as retarders, e.g., quarter-wave plates).

(*ii*) The density matrix of atoms idealized as two-level systems, that are in thermal equilibrium will be diagonal with respect to the basis given by the two energy eigenstates. Hence it appears natural to think of the atoms as being in such a state each. However, the mixture can be decomposed, e.g., also into superposition states (2.16) with the same moduli, following from Boltzmann statistics, of the coefficients c_1 and c_2. The coefficients differ, however, in their relative phases δ such that the average over $\exp(i\delta)$, and hence the nondiagonal elements of the density operator, vanish. (For instance, the subensembles contribute with equal weights w_k and the values of δ are distributed equidistantly over the interval from 0 to 2π.) So the whole ensemble can be considered as being composed of subensembles describing atoms that are coherently excited which might result in the presence of dipole oscillations. The induced dipole moments of all subsystems, however, add up to zero.

(*iii*) A similar situation can be found in laser theory. Since laser operation starts from spontaneous emission, the phase of the laser radiation is random. Hence the density operator of a single-mode laser field, as it is obtained from an analysis of the interaction between radiation and a laser medium, is diagonal with respect to the Fock basis (states of sharp photon numbers), the diagonal elements describing, to a good approximation, a Poisson photon distribution. So the density operator is readily decomposed into Fock states. Since the phase of radiation fields being in such a state is indeterminate (the expectation value of the electric field strength vanishes), this might arouse protests at least from experimentalists, since they know pretty well that laser radiation has an extraordinarily sharp phase. This contradiction between theory and reality is, however, only apparent. Owing to the ambiguity of the decomposition it is not justified to conclude from a possible decomposition that the individual systems are in the respective pure states. On the other hand, the density operator can also be decomposed into coherent states $|\alpha\rangle$ with fixed $|\alpha|$ and equidistantly distributed phases of α, similar to the atomic case discussed before. The latter decomposition actually meets our expectation; nevertheless, it is not compelling.

We should keep in mind that quantum mechanics describes only ensembles. So the density operator in question gives us no detailed information on the properties of the radiation field you have just produced by switching on your laser. Fortunately, to test the quantum mechanical predictions you need not operate, say one thousand lasers simultaneously; you can do as well by repeatedly switching on and off the single laser you could afford. Then forming averages over all those fields, you will find agreement with theory. In particular, since the phase of a laser field is, in fact, unpredictable, the average of the electric field strength will vanish.

Nevertheless, the situation is unsatisfactory. We have seen that quantum mechanics fails to match the classical description, based on the concept of 'objective reality', of the macroscopic world.

A peculiar feature of the quantum mechanical state concept shows up when we are considering a system that is composed of two (or more) interacting subsystems. We focus on the special case that observations are made on one subsystem only. Then the operators representing the observables act as unit operators in the Hilbert space of the unobserved subsystem. Hence the observed subsystem can be described by a reduced density operator that is obtained from the density operator for the total system by forming the trace over the subspace of the unobserved subsystem. Actually, this is similar to what we are accustomed to in classical statistics: from a distribution function for the variables of the total system we get a reduced distribution function for a subsystem by integrating over the variables of the nonobserved subsystem.

The reduced density operator normally describes a mixture. (The only exception is the case when the total system is in a pure state that factorizes into pure states for the subsystems.) This is even so when the total system is in a pure state! This statement is really puzzling. Actually, it is at variance with classical statistics: when a system is in a 'pure state' (a state in which all variables are sharp), the same necessarily applies to its subsystems.

Let me illustrate the quantum mechanical peculiarity under discussion using laser theory. When initially all atoms of the active medium are excited, the latter is in a pure state. The same holds true for the initial radiation field, which is in the vacuum state. The initial wavefunction of the total system is just the product of the wavefunctions for the laser medium and the field. Since the Schrödinger equation (2.8) transforms a pure state again into a pure state, the total system is still in a pure state when a laser field has actually built up. This pure state comes into play only when we study correlations between the medium and the field (for instance, the phases of the macroscopic dipole moment and the laser field are strongly correlated). However, when we take notice of the field only, it appears to us as being in a mixture. This sounds rather paradoxical, but it is an unavoidable consequence of the quantum mechanical formalism.

2.4 Wigner function

Unlike a classical distribution, which is a function of a full set of physical variables, the quantum mechanical wavefunction depends on half the variables only, e.g., the coordinates or the momenta. This is, in fact, not surprising in view of Heisenberg's uncertainty relation, which prevents us from assigning sharp values both to position and momentum. Nevertheless, in quantum theory one might imitate classical distributions by introducing so-called quasiprobability distributions that also depend on all variables. The pioneering work is due to Wigner, who defined what later was named the Wigner function or Wigner distribution as follows:

$$W(x,p) = (\pi\hbar)^{-1} \int_{-\infty}^{+\infty} \exp(2ipy/\hbar)\psi(x-y)\psi^*(x+y)dy \qquad (2.24)$$

or, more generally,

$$W(x,p) = (\pi\hbar)^{-1} \int_{-\infty}^{+\infty} \exp(2ipy/\hbar)\langle x-y|\hat{\rho}|x+y\rangle dy. \qquad (2.25)$$

Here, we have focused, for simplicity, on the one-dimensional case; x is the position and p the momentum of a particle. On the other hand, the variables x and p may also stand for the quadrature components of a single-mode radiation field (see Section 2.7.2). (In the latter case we have to drop $\hbar$ in (2.1) and consequently also in (2.24), (2.25) and (2.26)). On the other hand, from a given Wigner function the density matrix, in x representation, is readily obtained by inverting (2.25), as the

Fourier transform

$$\langle x | \hat{\rho} | x' \rangle = \int\limits_{-\infty}^{+\infty} \exp[\mathrm{i}(x - x')p/\hbar] W \left[\frac{1}{2}(x + x'), p \right] \mathrm{d}p. \qquad (2.26)$$

Hence it is clear that the Wigner function is a perfect substitute of the wavefunction or the density operator. It contains the full quantum mechanical information on a system.

Now, what makes the Wigner function attractive for the theorist? Wigner's own motivation was to use it for studying quantum corrections to classical statistical mechanics. Apart from this, the Wigner function has the great advantage that it can be directly visualized. In fact, unlike the complex wavefunction or the abstract density operator, it is a real function, and with the help of modern computer techniques graphic representations (at least for a one-dimensional system) are readily generated. It should be kept in mind, however, that the Wigner function, in contrast to a true probability distribution, has no direct physical meaning. This is illustrated, in particular, by the fact that it normally also takes negative values (the only exception being Gaussian states). In fact, domains of negativity can be associated with nonclassical features. The normalization condition the Wigner function has to fulfil,

$$\int\limits_{-\infty}^{+\infty} \int\limits_{-\infty}^{+\infty} W(x, p) \mathrm{d}x \mathrm{d}p = 1, \qquad (2.27)$$

tells us that positive values must dominate. It should be noticed that you cannot declare *any* real function satisfying (2.27) to be a Wigner function. Actually, the function has to obey rather sophisticated mathematical criteria. In this respect the Wigner function differs unfavourably from state vectors (wavefunctions or Hilbert space vectors), which are assumed to represent always a physical state that is possible, at least in principle.

Though it is not clear what we actually see, the pictorial representations of the Wigner function give us an immediate impression of the respective state (see, e.g., Fig. 2.1); for instance, we may recognize, at one glance, characteristic symmetries. Interestingly enough, for a single-mode radiation field an experimental scheme was even devised (Banaszek and Wódkiewicz, 1996) that allows one to determine the Wigner function from the photon statistics to be measured on the field after it has been mixed with a high-intensity coherent field. Varying the amplitude and the

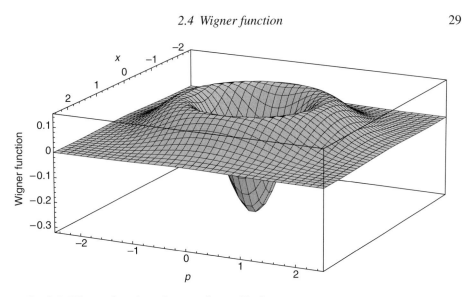

Fig. 2.1 Wigner function of a one-photon Fock state.

phase of this reference field, one finds the Wigner function of the original field point by point. Further, the Wigner function proved to be a very useful tool in the reconstruction of radiation field states: it is just the Wigner function that can be reconstructed directly from measured data (see Section 2.7.2).

Now we have to answer the question: when the Wigner function is no probability distribution, what is it then? Well, it is a clever theoretical construct on which the quantum mechanical description may be based (see, in particular, Schleich (2001)). The rules for extracting physical information from the Wigner function are as follows.

First, we can write the expectation value for any operator function $F(\hat{x}, \hat{p})$ that is *symmetrically ordered* with respect to the position operator $\hat{x}$ and the momentum operator $\hat{p}$, in the form of a classical average

$$\langle F(\hat{x}, \hat{p}) \rangle = \int_{-\infty}^{+\infty} \int_{-\infty}^{+\infty} W(x, p) f(x, p) \mathrm{d}x \mathrm{d}p. \tag{2.28}$$

Here, $f(x, p)$ is the function that originates from $F(\hat{x}, \hat{p})$ when the operators $\hat{x}$ and $\hat{p}$ are replaced by the corresponding c-numbers x and p. The drawback of this prodecure is the need for symmetrical ordering, which normally requires repeated use of the commutation relation (2.1). This makes the calculation often rather troublesome.

Second, we get the probability distributions for measuring x and p, respectively, by forming the marginals of the Wigner function

$$w(x) = \int\limits_{-\infty}^{+\infty} W(x,p)dp, \quad w(p) = \int\limits_{-\infty}^{+\infty} W(x,p)dx. \tag{2.29}$$

While the evolution of the Wigner function is governed by a rather intricate differential equation, in general, it becomes amazingly simple in the case of systems that are coupled via a quadratic interaction Hamiltonian. Then you can forget about quantum mechanics and calculate the time evolution *as if* the Wigner function were a classical distribution. The classical picture is well known: any point in the phase space of the total system, $P = (x_1, p_1; x_2, p_2; \ldots)$, will move along a certain trajectory, starting from some point $P^0 = (x_1^0, p_1^0; x_2^0, p_2^0; \ldots)$, and during this travel the initial value of the distribution at P^0 remains attached to the phase space point. So we have

$$W(P;t) = W(P^0;0). \tag{2.30}$$

What we have to do is to solve the classical equations of motion (which is easy, since they are linear in our case) and invert them. This gives us the initial values $x_1^0, p_1^0; \ldots$ as functions of $x_1(t), p_1(t); \ldots$, and those functions have to be inserted on the right-hand side of (2.30). The inverted classical solutions are linear functions too, and the time dependence is contained in the coefficients. It should be emphasized that (2.30) describes, in fact, *exactly* the quantum mechanical evolution of an *arbitrary* quantum state. Though we only needed to make simple classical calculations, we took properly into account the quantum mechanical subtleties that might be inherent already in the initial state, or evolve in time. The correspondence between the classical and the quantum mechanical description could not be closer!

From the simple procedure just described one benefits especially in quantum optics, where the arguments of the Wigner function have to be identified with the quadrature components of the electric field strength (see Section 2.7.2). In particular, the scheme applies to beamsplitting and optical parametric amplification. A beamsplitter (a partly transmitting mirror assumed to be lossless) can be used as an optical mixer. Then two light beams are separately sent onto the mirror under the same angle, however, from different sides (see Fig. 4.2). Each of them will undergo both reflection and transmission. In the given circumstances, the reflected (transmitted) beam emerging from the first input beam gets superposed with the transmitted (reflected) beam from the second input beam so that interference takes place. So in this experimental arrangement the beamsplitter has two input and (only) two output ports. Now, the point is that the variables corresponding to the ingoing waves, on the one hand, and the outgoing waves, on the other hand, are connected through a linear (unitary) transformation. Hence, the aforementioned simple scheme to solve the evolution problem in the Wigner formalism can be applied.

What is usually understood by beamsplitting is the special case in which only one field is sent onto the mirror. From the classical point of view, this means that you can forget about the second input port which is unused now. But you cannot do so in quantum theory. Notwithstanding the fact that no electromagnetic energy is coupled into the unused input port, quantum theory claims that there is actually an input field which, however, is in its vacuum state, i.e., a Fock state with zero photons. The Wigner function for this state is well known; it is simply a Gaussian with respect to the quadrature components x and p. In this way, the unused input port actually comes into play and one can argue along the same lines as in optical mixing. This gives us, in fact, a very elegant means of describing quantum mechanically the action of a beamsplitter on an input field that is in an *arbitrary* quantum state.

From the standpoint of classical physics, it is hard to understand that, though there is no energy present, there exist fluctuating electric and magnetic field strengths. Those so-called vacuum fluctuations enter the unused input port, thus acting as a noise source whose effect on the reflected and the transmitted light can actually be observed (see Section 3.5).

In optical parametric amplification based on three-wave interaction in a nonlinear crystal (see Section 4.4.2) we are also dealing with linear equations of motion, provided the pump wave is so strong that it can be approximated by a classical wave of constant amplitude and phase. So we may also in this case benefit from the Wigner formalism (see Section 4.5).

Finally, I would like to mention that different quasiprobability distributions, actually an infinite manifold of them, can be constructed from the Wigner function by smoothing it, that is, by convoluting it with a Gaussian whose width may be chosen arbitrarily. Among them of special interest is the so-called Husimi or Q function, which is defined as

$$Q(x,p) = \pi^{-1} \int\limits_{-\infty}^{+\infty} \int\limits_{-\infty}^{+\infty} W(x',p') \exp\{-[(x-x')^2 + (p-p')^2]\} dx' dp'. \quad (2.31)$$

Contrary to the Wigner function, the Q function is positive definite. Hence it is a true distribution function. Actually, we know physical situations in which the Q function is directly measurable (see Sections 3.5 and 4.5). Then the smoothing of the Wigner function, which wipes out finer details of the latter and thus leads to a loss of information, has a simple physical interpretation: it reflects the presence of noise.

2.5 How to describe a single system?

In the foregoing text it has been emphasized that the quantum mechanical state concept applies to ensembles. They allow us to testify the probabilistic quantum mechanical predictions of what will happen in measuring a given observable. In particular, statements saying that a certain observable has a sharp value can be verified on an ensemble. Consider, for example, an ensemble of photons that is asserted to be in the x polarization state. As a measuring apparatus we will use the device already mentioned in Section 2.2, namely a polarizing prism with a separate detector in each exit. Choosing the orientation of the prism such that its directions of transmittance coincide with the x and the y directions, respectively, the experimental criterion is that we will detect the photon *always* in the x polarization exit port, and never in the other. In fact, just one exceptional event – a response of the detector in the y polarization exit port – would suffice to falsify the assertion made. So we can say that physical parameters can be attributed to ensembles as objective properties not influenced by subjective factors. This viewpoint also finds strong support from the fact that the quantum state can actually be reconstructed from a variety of measurements performed on the ensemble (see Section 2.7).

But what about single systems? Does it make sense to describe an individual system by a quantum state? In fact, one is tempted to say that a single system is in a definite quantum state when it is known to be a member of an ensemble being in this state. This amounts to having precise information on the preparation process. Practically, this means that we must know, in some detail, the apparatus used for preparation. In the case of strongly correlated systems, such as the entangled photon pairs described in Section 2.6.5, the information on the wavefunction of the total system can be combined with that obtained from a measurement on one subsystem (which formally gives rise to a reduction of the aforementioned wavefunction) to yield the wavefunction of the unobserved subsystem.

We are facing, however, serious difficulties. Owing to the intrinsically probabilistic nature of the microscopic world, we can only specify, on principle, probabilities for certain events to occur. Such probabilities can never be checked on a single system – actually, the situation resembles that when doctors tell us that we have, say a 60 per cent chance of surviving a serious disease. Probabilistic predictions can only be used to guess what will happen. They may help us to win a bet. However, there is an exception, namely the measurement of just that observable of which the state is an eigenstate. Then we can predict the outcome of the measurement *with certainty*. We can be sure that the eigenvalue will always be measured. However, even such a fully certain prediction can never be verified experimentally. Though a measurement of the observable in question will confirm the prediction, you can never be sure that it was not merely by accident. What can be concluded from the measurement is only that the system was previously in a superposition state with a nonzero admixture of the measured state. Actually, we will tentatively conclude from the measurement that this admixture is noticeable, since we do not believe that we were just witnessing an extremely unlikely event. But this belongs to the category of guessing.

From what has been said it becomes obvious that you can never get reliable information on the state of a single system from measurement. Only when you have prepared the system, or witnessed this process, or have a good friend who tells you about it, will you have knowledge of the state of the system. So, if you are lucky, you may have this information, but you cannot gain it from Nature itself. So the concept of a quantum state describing an individual system is dubious. Searching for such a description, we were actually misled by the deterministic behaviour we find in classical physics. From that viewpoint it is, in fact, hard to understand that a microscopic system follows intrinsically statistical, and hence indeterministic, laws which make its future unforeseeable. What we really feel as disquieting is that we cannot find out, in principle, a cause responsible for a particular event, e.g., an absorption of a single photon by just one out of, say, 10^{12} atoms.

Finally, it should be emphasized that you can never check the presence of indeterminacy on a single system. This is so, of course, in classical physics, where all variables are supposed to have sharp (though usually unknown) values in any individual case. Equally, quantum mechanical uncertainties in the sense of classical statistics (as they appear in the description of mixed quantum states) cannot be observed on a single quantum system, and the same also holds true for intrinsic uncertainty. Naturally, indeterminacy of any kind can be detected only by repeated measurements. This means that we have to deal with an ensemble of (equally prepared) systems. Uncertainty is indicated by the experimental finding that the measured values of the respective variable differ. This, however, does not tell us whether the observed uncertainty is of statistical (in the sense of classical statistics) or intrinsic nature. To verify the presence of intrinsic indeterminacy, one actually has to carry out a suitable *indirect* measurement. For instance, the intrinsic indeterminacy of the path a photon takes in an interferometer gives rise – according to theory – to interference. So, when an interference pattern is actually observed, we will conclude that the photon's path must have been (intrinsically) indeterminate.

2.6 State preparation

As in classical physics, in quantum mechanics the description or prediction of an actual experiment also requires the knowledge of the initial state. So it is important to answer the question: 'How can systems be prepared in a definite quantum state?' Astonishingly enough (and fortunately for quantum physicists), macroscopic apparatuses are suited to do this job. Naturally, one can assign a quantum state to the systems exiting from the device only when use is made of basic quantum concepts such as quantization and eigenstates of observables. Often one cannot even do without detailed quantum mechanical calculations. While state preparation is a simple thing for the theorist – choose an appropriate observable, make a measurement (in thought!), and you can be sure that the systems corresponding to a certain eigenvalue are in the respective eigenstate – it is a challenge to experimentalists. Not only experimental skill and carefulness, but also creative ideas (apart from money) are required.

In quantum optics one can fall back on well-tried techniques developed, over long periods, in classical optics. Indeed, the first step to prepare photons in a definite quantum state is to alter, in a desired manner, the spatial, spectral and polarization properties of light emitted from either a conventional (thermal) source or a laser, using well-known optical instruments such as diaphragms, mirrors, lenses or more sophisticated optical imaging systems, frequency filters and polarizers. (Temporal shaping is provided by lasers emitting sequences of short or ultrashort pulses.) The second step is to attenuate this light by letting it pass through an absorber of suitable absorptivity and dimensions. What is thus prepared, at best, is an ensemble of photons that arrive at a detector one after the other, however at irregular (and unpredictable) instants of time.

What are generally prepared with the help of realistic devices are ensembles. For a long time the common opinion was that one cannot do better. Schrödinger wrote in 1952: 'In gedanken experiments we sometimes assume that we do (experiments with single electrons and atoms); this invariably entails ridiculous consequences.' However, Schrödinger underrated the ingenuity and skill of experimentalists. Actually, researchers succeeded in making individual systems an object of observation. A decisive step is, of course, the preparation of single systems. This was achieved, in particular, for ions and photons (see Sections 2.6.3 and 2.6.5).

To give you an impression of the state of the art in quantum state preparation, I will shortly describe a few important experimental techniques.

2.6.1 Optical cooling

For experiments in atom optics low-velocity atoms are needed. Since atomic beams are produced by heating (the atoms escape, through a small hole, from an oven) their velocity is rather high. So it becomes necessary to slow them down rather drastically. This can be done very effectively by exploiting the radiation pressure exerted by strong laser radiation. The basic mechanism is the resonant absorption of a laser photon by an atom, followed by spontaneous emission. Resonance means that the Doppler-shifted laser frequency, $v' = v(1 + v/c)$ (v frequency in the laboratory system and c velocity of light), which is actually 'seen' by an atom moving with velocity v opposite to the laser beam, coincides with the atomic transition frequency. From momentum conservation, the atom suffers a recoil in the absorption process, whereas it remains unaffected, on the average, in spontaneous emission, since the photon is emitted in a random direction. Thus the atom is slowed down a little. The point is, however, that the mentioned cycle (which brings the atom back into the initial state) can be repeated many thousand times, which results in a drastic 'cooling' of the atoms. Since, with decreasing velocity, the atoms get out of resonance for fixed v, it is necessary to tune the laser frequency properly in the course of time, in order to maintain resonance. This can be achieved by employing 'chirped' laser pulses. (One speaks of chirping when the mid-frequency of the pulse varies monotonically with time, so that different parts of the pulse have different frequencies.)

2.6.2 Magneto-optical trap

The light pressure can be used also as a restoring force, which allows one to trap atoms. The atoms move in a static magnetic field that increases with the distance from a centre. This field causes a Zeeman splitting of the upper atomic level. This splitting, being proportional to the magnetic field strength, varies while the atom is moving. Shining circularly polarized laser light at a properly chosen frequency on the atoms, one finds that an atom becomes resonant (note that the laser frequency 'seen' by the atoms is Doppler-shifted) just when it is moving away from the centre, opposite to the laser wave, and hence suffers a recoil due to absorption. A trap is realized experimentally with two coils with opposing currents, which produce a magnetic quadrupole field. Superimposed is a radiation field consisting of three mutually orthogonal pairs of counter-propagating, oppositely circularly polarized laser beams. Turning off the trapping magnet and the optical fields, one gets free atoms that are virtually at rest. They were used, for example, to produce an 'atomic fountain': a resonant vertical laser beam launches them on ballistic trajectories, i.e., they move upwards and eventually fall back under the influence of gravity. With their free fall, the atoms fulfil the keenest dreams of a spectroscopist.

2.6.3 Paul trap

Ions can be trapped in an electric radio-frequency quadrupole field. It prevents captured ions from escaping. Actually, a static electric quadrupole field would not do; the trick is to invert the ion's motion periodically by periodically changing the sign of the voltage applied to the electrodes (two ring electrodes and two end caps).

Amazingly, not only can such a trap (which is named after Wolfgang Paul) be filled with just one ion, applying the optical cooling technique, but one can also store it over unlimited time intervals. The cooling is provided by a laser beam that damps, via radiation pressure, the vibrational motion of the ion. To this end, the laser frequency is chosen such that the ion becomes resonant just when it is moving counter to the field. Such a trapped ion is certainly an almost ideal starting point for further investigations. For instance, one can study resonance fluorescence from just one ion (cf. Section 3.3.1). Schrödinger's scepticism could not have been disproved more convincingly!

2.6.4 Coherent excitation

It is well known that atoms (as well as molecules) can be excited through absorption of light. Usually, stimulated emission and relaxation processes (such as spontaneous emission and collisions) compete with excitation, which brings the atoms into a thermal equilibrium state, with respect to the levels involved in the transition. This means that one can never attain population inversion, the number of atoms in the upper level being always smaller than in the lower one. This holds true, however, only for long excitation pulses, in particular for stationary fields. In fact, the situation changes drastically when the excitation process is shorter than the relaxation times. Then relaxation processes do not come into play, and the dynamics become coherent. Actually, the field (assumed to be resonant with an atomic transition starting from the ground state) generates an atomic superposition state of the form of (2.16). In contrast to the thermal equilibrium state, this is a pure state, which means that the atoms form a homogeneous ensemble.

The probability $w_2(t)$, given by $|c_2(t)|^2$, of finding an atom in the excited state proves to be an oscillating function of time,

$$w_2 = \sin^2\left(\frac{\Omega}{2}t\right), \qquad \Omega = \frac{D_{12}A}{h} \tag{2.32}$$

(for more details see Section 3.3.3). Here, we have focused on a dipole transition; D_{12} is the transition matrix element (assumed positive), $D_{12} = \langle E_1 | ez | E_2 \rangle$, where the field has been assumed to be polarized in the z direction, and A is the amplitude (assumed constant) of the electric field strength residing at the atomic position. The parameter Ω has been named the Rabi frequency, and the oscillation (2.32) is known as the Rabi oscillation.

Starting from zero at $t = 0$, w_2 grows continually until it reaches unity, and afterwards it decreases (owing to stimulated emission), reaches zero and starts the cycle again. So it becomes possible to bring *all* atoms simultaneously into the excited state. To keep them in this distinguished state, care must be taken that the interaction stops when the state is reached. This is a requirement of the pulse form. A theoretical analysis shows that the pulse area defined as

$$\theta = \int_{-\infty}^{+\infty} \frac{D_{12}\varepsilon(t)}{\hbar} dt \tag{2.33}$$

determines the final state of the atoms. Here, $\varepsilon(t)$ is the pulse envelope.

For the special value $\theta = \pi$ (accordingly, one speaks of a π pulse) the atoms will end up in the excited state, whereas for a $\pi/2$ pulse ($\theta = \pi/2$) we have $|c_1| = |c_2|$ in the final state, which implies that the induced dipole moment (expressed by the expectation value of the dipole operator ez) oscillates with maximum amplitude. Fortunately, modern laser technique allows us to tailor ultrashort pulses, so that we can manipulate atoms at will.

2.6.5 Entangled photon pairs

A stroke of luck for quantum optics is spontaneous parametric down-conversion, which produces photon pairs with marvellous properties. Let me dwell a little on this striking phenomenon. It is based on three-wave interaction mediated by a nonlinear crystal. This process is well known in (classical!) nonlinear optics. Here, two intense coherent running waves (in practice, laser waves) at frequencies ν_p and ν_s ($\nu_p > \nu_s$) and with wave vectors k_p and k_s, respectively, produce in the nonlinear crystal a macroscopic polarization (that is the sum over the induced atomic dipole moments contained in a unit volume) that oscillates with the frequency difference $\nu_i = \nu_p - \nu_s$. This polarization is actually a running wave too, its wave vector being given by $k_p - k_s$. The polarization wave, in turn, acts as a source of radiation. It can generate radiation at frequency ν_i which is propagating with a wave vector k_i. (Note that ν_i and k_i are bound to satisfy the dispersion relation for the crystal.) This generation process, named parametric down-conversion, is, however, effective only when k_i coincides with $k_p - k_s$. This so-called phase-matching condition is easily understood when noticing that the wave grows because of the work done by the polarization on it. Obviously, a fixed phase relation must exist between the two waves over the whole crystal, since otherwise amplification of the light wave would be followed by attenuation during its propagation through the crystal. So we have learned that the following conditions must be fulfilled simultaneously:

$$\omega_p = \omega_s + \omega_i \ (\omega = 2\pi\nu), \tag{2.34}$$

$$k_p = k_s + k_i. \tag{2.35}$$

While the crystal does not exchange energy with the fields – it acts like a catalyst – the newly generated wave gets its energy from the incident wave at the higher frequency ν_p (this wave is therefore named 'pump wave' which explains the label 'p'). A closer inspection shows that the incident wave at ν_s also receives a share of the pump energy, thus becoming amplified as the wave at ν_i develops. Hence the device can be used also to amplify a weak wave at ν_s, which is then called the signal. The third wave at ν_i is necessarily generated too; however, it is of no use and is, hence, called 'idler'. The whole device is known as an optical parametric amplifier. Since energy at a higher frequency, ν_p, is converted into energies at lower frequencies, ν_s and ν_i, one also speaks of parametric down-conversion.

It is important to note that the relations (2.34) and (2.35) are readily interpreted in the photon picture. After their multiplication by $\hbar$, the first says that a pump photon is 'split' into a signal photon and an idler photon and hence expresses the energy conservation law. The second equation, on the other hand, is a conservation law for the momenta of the photons being involved in the interaction.

Until now, we have assumed both the pump and the signal wave to be sent into the crystal, so that parametric down-conversion is an induced process. Now we know from atomic emission that induced and spontaeous emission have the same physical root. Accordingly, when induced emission can take place, spontaneous emission must also be possible, and vice versa. This general rule also applies to the parametric three-wave interaction under discussion. So when we send only the pump wave – in practice, a laser wave – into the crystal, photon pairs consisting of a signal and an idler photon will be emitted spontaneously. The possible frequencies and propagation directions are determined by (2.34) and (2.35). When the pump wave is strong enough, spontaneously emitted photon pairs will become the seed of intense waves, and we end up with an optical parametric oscillator which, in fact, is a tunable source of laser-like light. (Frequency tuning is achieved by rotating the crystal or changing its temperature.)

When the pump intensity is properly chosen, the regime of spontaneous parametric down-conversion is maintained: single photon pairs emerge from the crystal one after the other. Actually, they are invaluable objects for experimental studies. Their characteristics depend on how the phase-matching condition (2.35), in conjunction with the energy conservation law (2.34), is fulfilled.

In the so-called type I interaction, both the signal and the idler photon are linearly polarized in the same direction. However, the two photons are normally of different colour, and they propagate in different directions. When the pump wave, assumed to be approximately a monochromatic plane wave, propagates parallel to the optical axis of the crystal, the propagation directions lie on cones (with their axes parallel to the pump wave direction) corresponding to different colours. However, the two photons are strongly correlated, as becomes obvious from (2.34) and (2.35).

As long as no measurement is made on them, both their frequencies and propagation directions are uncertain. The situation changes drastically, however, when one of the photons is subjected to a measurement. Placing an aperture at an appropriate position, we can select photons – let us call them signal photons – whose frequency and propagation direction are well defined. (When necessary, their bandwidth can be further reduced with the help of a frequency filter.) Then the correlations imply that the corresponding *unobserved* idler photons will also have a definite frequency and propagate in a fixed direction. The frequency and the wave vector are readily read from (2.34) and (2.35) after insertion of the corresponding values for the pump wave and the observed photon. It should be emphasized that we are dealing with a real selection process. Most of the signal photons will be lost (absorbed by the aperture).

In particular, the experimental set-up allows us to perform true single-photon experiments. Placing a detector behind the pinhole, its response gives us the information that a second photon with known frequency and propagation direction is present. Moreover, since the signal and the idler photon are generated by the splitting of a pump photon, they must be correlated very strongly in time: they exit from the crystal with no delay between them. So spontaneous parametric down-conversion is ideally suited for state preparation of single photons.

In appropriate crystals (beta-barium borate proved to be a good choice) a different type of interaction, named type II, can be realized experimentally. Here, the polarization directions of signal and idler are orthogonal. This gives us an opportunity to produce polarization correlations of a peculiar kind. We focus on the degenerate case ($\nu_s = \nu_i$). Then the propagation directions of the signal and the idler photon each lie on a cone. The cones correspond to ordinary and extraordinary beams in the crystal, respectively, which are mutually orthogonally polarized, say in the x and y directions. When the propagation direction of the pump wave is tilted with respect to the optical axis of the crystal, the two cones become separated, their axes being differently oriented. Actually, the propagation direction of the pump can be chosen such that the two cones intersect (see Fig. 2.2). This gives us two distinguished propagation directions, which we will label 1 and 2. A photon propagating in such a direction (experimentally, this requires selection with the help of an aperture) does not 'know' to which cone it belongs, and hence, what its polarization is. Now, when the signal photon is propagating, say in direction 1, its counterpart, the idler photon, is propagating in direction 2. The mentioned polarization uncertainty is quantum mechanically described by a superposition state of the form

$$|\psi\rangle = \frac{1}{\sqrt{2}}(|x\rangle_1 |y\rangle_2 + e^{i\alpha} |y\rangle_1 |x\rangle_2), \qquad (2.36)$$

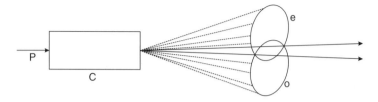

Fig. 2.2 Type II parametric down-conversion. The propagation directions of the two simultaneously emitted photons each lie on a separate cone. P = pump wave; C = nonlinear crystal; o = ordinary (vertical) polarization; e = extraordinary (horizontal) polarization.

where $|x\rangle_1$ denotes a photon polarized in the x direction and propagating in direction 1, etc. The relative phase α is determined by the crystal parameters. By placing a birefringent phase shifter in one of the paths, it can be set at will. In particular, it can be made 0 or π.

The strong correlations between variables of different systems as described, for example, by the wavefunction (2.36), or expressed by the conditions (2.34) and (2.35), were baptized 'entanglement' by Schrödinger. Its characteristic feature is that the variables are indeterminate; however, a measurement on one subsystem uniquely fixes the corresponding variable of the other system. Actually, entanglement is a fundamental quantum mechanical concept, as will be explained in Section 4.3.

2.7 State reconstruction

2.7.1 Phase retrieval

In view of the fundamental role the quantum state plays in quantum theory, one is led to ask whether it might be determined from measurements, thus demonstrating its 'objective' meaning. A typical problem is the reconstruction of the wavefunction $\psi(x)$ for a particle. In this case, the first step offers itself: the modulus of $\psi(x)$ can readily be determined from position measurements. In fact, since $|\psi(x)|^2\,d\tau$ is the probability of detecting the particle in a volume element $d\tau$ located at x, $|\psi(x)|^2$ can be found from the outcomes of a large number of position measurements. What remains undetermined yet, and hence must be inferred from different measurements, is the (space-dependent) phase of the wavefunction. This task is known as phase retrieval. One will guess that the missing information can be gained from measuring the conjugate variable, that is, the momentum. Since the probability distribution for the momentum is given by the squared modulus of the Fourier transform of $\psi(x)$, we face the problem of reconstructing a complex function from its modulus and the modulus of its Fourier transform.

It is interesting to note that just the same problem arises in classical optics, where ψ is the complex amplitude of the electric field strength. In particular, in microscopy (optical as well as electron microscopy) one wants to reconstruct the wavefunction in the image plane, $\psi_i(x, y)$, from the intensity distributions in the image plane and the exit pupil, $|\psi_i(x, y)|^2$ and $|\psi_e(\xi, \eta)|^2$. Since $\psi_e(\xi, \eta)$ is the Fourier transform of $\psi_i(x, y)$, those distributions are, in fact, related in the same manner as those for position and momentum in quantum theory.

Unfortunately, the mathematical problem in question is of a nonlinear type and hence does not allow for an analytic solution. It has been discussed especially in the context of electron microscopy. Various approximative algorithms were devised, notably the Gerchberg–Saxton algorithm (Gerchberg and Saxton, 1972), which is based on an iterative scheme.

2.7.2 Tomography

In quantum optics, a breakthrough came from the observation that the reconstruction problem for the state of a single-mode field can be cast in a form that is known from X-ray, or computer, tomography which has become a powerful tool in medicine. Here, the aim is to construct three-dimensional pictures of areas within the human body from measurements of the damping suffered by thin X-ray beams. While the measured intensity loss yields direct information on the *total* absorbance of the tissue – the integral over the local absorbance along the beam – what one wishes to know, however, is the *local* absorbance whose spatial distribution will give us the desired information, e.g., on the shape and the location of an organ or a tumour. This goal is achieved by sending X-rays through the body not only in one, but in many directions. The main task is to determine the absorbance distribution $F(x, y)$ in a given cross-sectional slice. A three-dimensional picture is then readily obtained by composing the slices. A chosen slice of the body is scanned by (*i*) parallel displacement of a beam that traverses the slice parallel to the cross-section, and (*ii*) repetition of this procedure with beams rotated by an angle Θ (with the rotation axis normal to the slice). Any run (*i*) yields a marginal, dependent on Θ, of the absorbance distribution

$$f_\Theta(x_\Theta) = \int_{-\infty}^{+\infty} F_\Theta(x_\Theta, y_\Theta) dy_\Theta. \qquad (2.37)$$

Here, the coordinates x_Θ and y_Θ refer to a basis that is adapted to the direction of the beam rotated by Θ, and hence rotated by Θ with respect to the x, y basis chosen primarily. So we have

$$x_\Theta = x \cos \Theta + y \sin \Theta,$$
$$y_\Theta = -x \sin \Theta + y \cos \Theta \qquad (-\pi/2 \le \Theta \le \pi/2).$$

(2.38)

The function $F_\Theta(x_\Theta, y_\Theta)$ arises from rewriting $F(x, y)$ as a function of x_Θ and y_Θ, $F_\Theta(x_\Theta, y_\Theta) = F(x, y)$.

Now, the decisive question is: how can we, if at all, reconstruct the function $F(x, y)$ from its marginals? The answer was given as early as 1917 by the mathematician J. Radon, who found an elegant solution that became known as inverse Radon transformation:

$$F(x, y) = \frac{1}{(2\pi)^2} \int\limits_{-\infty}^{+\infty} \int\limits_{-\pi/2}^{+\pi/2} \int\limits_{-\infty}^{+\infty} f_\Theta(\eta) \, |\xi| \, \exp\{i\xi(x \cos \Theta + y \sin \Theta - \eta)\} d\eta d\Theta d\xi.$$

(2.39)

We learn from this formula that the marginals for virtually all angles Θ within a π-interval are required. Needless to say, for practical purposes, one has to replace the integral over Θ by a sum over suitably selected discrete values of Θ.

So far so good. But what has this to do with quantum mechanics? Actually, the key that allows us to make use of tomography is provided by the Wigner function. The latter might be reconstructed, via the inverse Radon transform (2.39), from its marginals (cf. Leonhardt (1997)). But how to measure them? Fortunately, there is a physical system for which this can be done easily, namely a single-mode radiation field. As we will see in Section 3.2, optical homodyning does this job. So let me dwell a little on the quantum mechanical description of single-mode fields.

We start from the observation that such a system is formally fully equivalent to a harmonic oscillator. The analogues of position and momentum (of a material harmonic oscillator such as an elastically bound electron) are the so-called quadrature components of the electric field strength. They come into play when the latter is written in real terms. Specializing to a running plane wave, we may write the electric field strength as

$$E(z, t) = E_0[x \cos(\omega t - kz) + p \sin(\omega t - kz)],$$

(2.40)

where E_0 is a normalization factor. In classical optics, x and p are real canonically conjugate variables. Hence in quantum optics they are represented by Hermitian operators $\hat{x}$ and $\hat{p}$ that satisfy, in convenient normalization, the commutation relation

$$[\hat{x}, \hat{p}] = i\mathbf{1}.$$

(2.41)

The operators $\hat{x}$ and $\hat{p}$ are related to the familiar photon creation and annihilation operators $\hat{a}^\dagger$, $\hat{a}$ (see Section 8.1) in the form

$$\hat{x} = \frac{1}{\sqrt{2}}(\hat{a}^\dagger + \hat{a}), \quad \hat{p} = \frac{i}{\sqrt{2}}(\hat{a}^\dagger - \hat{a}). \tag{2.42}$$

Now, we learned in Section 2.4 that the marginal $w(x) = \int\limits_{-\infty}^{+\infty} W(x,p)dp$ is the probability distribution for the observable $\hat{x}$. Clearly, this rule applies also to the 'rotated' marginal $w_\Theta(x_\Theta)$ of the Wigner function, formed after the example (2.37). So what we have to do according to (2.38) is to measure the observable

$$\hat{x}_\Theta = \hat{x}\cos\Theta + \hat{p}\sin\Theta \tag{2.43}$$

for different values of Θ. That this is not only possible but even feasible will be explained in Section 3.2. In particular, the required rotation proves to be very simple: it is accomplished by varying the relative phase between the signal and the reference field it is mixed with. The pioneering experimental work was done by the Raymer group (Smithey, Beck and Raymer, 1993), who succeeded in reconstructing the Wigner function for squeezed light and the electromagnetic vacuum.

In fact, the tomographic scheme is not confined to quantum optics. It can be applied also to material systems, e.g., monoenergetic atomic beams. In this case one might be interested in the wavefunction describing the transverse motion of a particle that has passed a diffraction device, for instance a double slit, and is then freely propagating. We obtain the 'rotated' marginals by letting the time work for us. They are related, in a simple way, to the position distributions that correspond to different times and hence can be observed in transverse planes located at different distances from the diffraction device. Precisely speaking, as the evolution time can take only positive values, only half of the full π interval can be covered in this way. Actually, access to the remaining rotation angles is provided by using atom-optical elements such as lenses.

It is noteworthy that the concept of the Wigner function can be applied to classical optics too. Then, the Wigner function is defined as the Fourier transform of the cross-correlation function considered in a transverse plane of a quasimonochromatic, partially coherent, linearly polarized light beam such as a realistic laser beam. It can be reconstructed along just the same lines as in atom optics, namely from measurements of transverse intensity distributions in different planes, both under free-propagation conditions and in the presence of a focusing lens.

2.7.3 Reconstruction of the density matrix

In Section 2.4 it was mentioned that the Wigner function comprises the full information on a quantum system. In making quantum mechanical calculations we will, however, usually prefer to deal with density operators (with the wavefunction as a special case). Making use of (2.26) we can pass from the Wigner function to the density operator in x representation. Then we choose a different, more convenient basis in Hilbert space. This requires the density operator to be subjected to a certain unitary transformation.

Actually, the detour via the Wigner function is superfluous. It was shown that the density matrix element $\langle a | \hat{\rho} | a' \rangle$ with respect to any basis $\{|a\rangle\}$ can be obtained *directly* from the measured data with the help of the following formula (Leonhardt, 1997):

$$\langle a' | \hat{\rho} | a \rangle = \int\limits_{-\pi/2}^{\pi/2} \int\limits_{-\infty}^{+\infty} w_{\Theta}(x) F_{a'a}(x, \Theta) \, dx \, d\Theta, \tag{2.44}$$

where $w_{\Theta}(x_{\Theta})$ is the probability distribution for the observable (2.43), as before. Analytic expressions for the kernel $F_{a'a}(x, \Theta)$ were derived for the coherent-state and the photon-number-state basis (often called the Fock basis) which is favoured in quantum optics.

Equation (2.44) tells us that the matrix elements we are looking for are simply the averages of known 'pattern functions' $F_{a'a}(x, \Theta)$ with the measured distribution functions w_{Θ} as weight functions. Hence the density matrix elements differ favourably from the Wigner function in that they allow us to apply the sampling technique.

3

Measurement

In Section 1.2 basic measuring schemes were described that utilize multiplication processes to produce a macroscopic signal, which, in fact, is an indispensable condition for any measuring apparatus to meet. Actually, we can learn more from the detection of a particle than that it was present in a certain region of space. This is so when the particle is acted upon by a force that depends on the quantum state. This makes it possible to split a particle beam into separate beams corresponding to different quantum states. The measurement is completed by detecting the particle in one beam or the other. In this way the quantum state is measured.

3.1 Spin measurement

3.1.1 Stern–Gerlach experiment

A scheme of the aforementioned type was realized experimentally as early as 1921 by O. Stern and W. Gerlach. Their experiment became famous as the first demonstration of spin quantization. It relies on the fact, well known from classical electrodynamics, that a magnetic momentum μ, when placed in a static magnetic field B (magnetic induction), acquires the potential energy $-\mu B$, or $-\mu_x B$ when we denote the field direction as the x direction. In the Stern–Gerlach experiment the magnetic field is actually inhomogeneous, $B = B(x)$. The inhomogeneity is produced by giving one of the magnetic poles such a form that it narrows, like a wedge, to a sharp edge (see Fig. 3.1). In the inhomogeneous field the potential energy becomes space-dependent. As a result, in the x direction a force acts upon the magnetic momentum that is given by $\mu_x dB/dx$. This force causes a beam, traversing the magnetic field, of atoms having magnetic momenta to split in dependence on the projections of the momenta on the field direction, and hence on their orientation. So the apparatus is suited to measure μ_x, and since the magnetic momentum is proportional to the spin, also the spin component in the x direction.

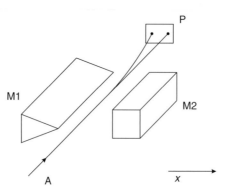

Fig. 3.1 Stern–Gerlach experiment. A = atomic beam; M1, M2 = magnetic poles; P = photographic plate.

Stern and Gerlach worked with silver atoms which possess an unpaired electron giving rise to the presence of a magnetic momentum $\mu = \mu_B \hbar s$, where s is the electron spin (in units of $\hbar$) and μ_B is a constant known as the Bohr magneton. Since the atoms emerge from an oven, their spins will be randomly orientated. Therefore, from a classical point of view one will expect that the values of s_x occurring in the beam will be uniformly distributed between $-l$ and l, where l is the length of the spin vector. Hence the beam will diverge. What was actually observed, however, was a distinct splitting of the beam into two. In fact, this is just what quantum theory predicts: generally, spins are characterized by quantum numbers J that are either integers or half-integers. For a spin with quantum number J the square of the spin operator has the eigenvalue $J(J+1)$, and the spin component with respect to any given direction has $2J + 1$ eigenstates with eigenvalues $-J, -J + 1, \ldots, J$. Specializing to the electron spin, we have $J = s = \frac{1}{2}$ and, in particular, $s_x = -\frac{1}{2}, \frac{1}{2}$. This explains the outcome of the Stern–Gerlach experiment, when we interpret it as a measurement of the component s_x of the electron spin.

This point needs a more thorough discussion. The result of the Stern–Gerlach experiment is easily understood when the spins are assumed to be *initially* orientated such that either $s_x = -\frac{1}{2}$ or $s_x = \frac{1}{2}$. Then the magnetic field acts on them just as in classical mechanics, it simply sorts the atoms according to their spin components s_x. This interpretation is actually compatible with the quantum mechanical description: since there is no preference for any spin direction, the state of the spins has to be described by a density matrix that is a multiple of the unity matrix with respect to *any* basis in spin space. We are free to choose the basis to be given by the two eigenstates of the spin component in the x direction, $\left|s_x = -\frac{1}{2}\right\rangle$ and $\left|s_x = \frac{1}{2}\right\rangle$, which allows us to conceive the spin ensemble to be composed of two subensembles with $s_x = -\frac{1}{2}$ and $\frac{1}{2}$, respectively. In this picture one misses the specific quantum feature of the measuring process.

However, we can do more. We may send one of the output beams of the Stern–Gerlach apparatus, say that prepared in the eigenstate $\left| s_x = \frac{1}{2} \right\rangle$, through a second Stern–Gerlach apparatus, with its magnet, however, rotated by an angle Θ, say $\Theta = \pi/2$. It turns out that the beam becomes split again, now into two beams corresponding to sharp values of the y component of the spin, $s_y = -\frac{1}{2}$ and $s_y = \frac{1}{2}$. This outcome is readily understood as a consequence of the linearity of the Schrödinger equation. In fact, the initial state $\left| s_x = \frac{1}{2} \right\rangle$ can be written as a superposition of the states $\left| s_y = -\frac{1}{2} \right\rangle$ and $\left| s_y = \frac{1}{2} \right\rangle$. Owing to the linearity of the evolution this superposition is preserved in the interaction, however atoms with spins in one or the other of the two eigenstates of s_y become spatially separated. This is really a typical quantum mechanical measurement: at the beginning, the y component of the spin is intrinsically uncertain, and as a result of the measuring process it acquires a sharp value. (From the formal point of view, the famous reduction of the wavefunction takes place.)

I would like to emphasize once again that the measurement on an atom is completed only when the atom is ultimately detected in one of the two beams. In this way the measurement result is read out. Unfortunately, the atom gets lost in this process! On the other hand, when the atoms are not detected, the splitting of the beam can be reversed, thus reproducing the original state $\left| s_x = \frac{1}{2} \right\rangle$, at least in principle. However, the detection can be postponed until after a second experiment has been performed. This is what we actually did in the experiment with crossed Stern–Gerlach magnets. This procedure amounts to simply ignoring the beam that we are not interested in.

It is interesting to note that there exists an electric analogue to the Stern–Gerlach experiment. In fact, when we are dealing with molecules whose energy levels are dependent on the electric field strength of a (static) external field, those molecules will experience state-dependent forces in an inhomogeneous electric field. In ammonia molecules we encounter the special situation that the upper level of the inversion doublet is shifted upwards with increasing electric field strength, while the lower level is shifted downwards. Hence differently excited molecules are subjected to opposite forces in an inhomogeneous field. Gordon, Zeiger and Townes (1954) had the good idea of exploiting this effect for creating a population inversion (this means, opposite to Boltzmann's distribution, more molecules are in the upper level than in the lower), which is a prerequisite of maser (and laser) operation. Letting a beam of ammonia molecules pass through a quadrupole electric field, the researchers focused the excited molecules so that they could enter a properly tuned microwave cavity. So the first maser was born.

3.1.2 Polarization measurement

The measurement of linear polarization on single photons already described in Section 2.2 is similar to the measurement of electronic spin components. A polarizing prism is the analogue of a Stern–Gerlach magnet. With its help, the polarizations in two orthogonal directions, say the x and y directions, are readily measured. In any single experiment, the photon will be found to be x or y polarized. However, we may rotate the prism, thus asking the photon whether it is x' or y' polarized. Furthermore, placing a quarter-wave plate in front of the prism, we can measure left- and right-handed circular polarization.

Nevertheless, the analogy is not perfect. We are not dealing with the total spin of the photon but with its projection onto the propagation direction. The eigenstates of this spin component are left- and right-handed circular polarization states with eigenvalues 1 and -1, respectively. The spin of linear polarization states, on the other hand, is uncertain. In fact, it is well known from classical optics that linearly polarized light can be thought of as being a superposition of left- and right-handed circular light, and the same holds true for the quantum mechanical polarization states of a single photon.

In the following I will describe some modern measuring techniques that proved to be relevant in quantum optics.

3.2 Balanced homodyne detection

While the idea of measuring the electric field strength of an optical field is illusory – no detector is conceivable that could follow, say, 10^{14} oscillations per second – the quadrature components can, in fact, be measured with the help of an ingenious technique. The basic idea is to mix the microscopic field under investigation (the signal) with a strong coherent field of equal frequency (in practice, a laser field), the so-called local oscillator. The latter acts like a macroscopic probe on the signal. For mixing, a beamsplitter with 50 per cent transmittivity and reflectivity is used (see Fig. 3.2). It produces a superposition of the transmitted signal and the reflected local oscillator in one output port, and of the reflected signal and the transmitted local oscillator in the other. More precisely speaking, the contributions from the signal and the local oscillator, respectively, are added in one port and subtracted in the other. Now, letting the output beams impinge on separate photodiodes yields two macroscopic electric currents. They are clearly dominated by the local oscillator. The trick is to subtract them. Then the current components due to the local oscillator

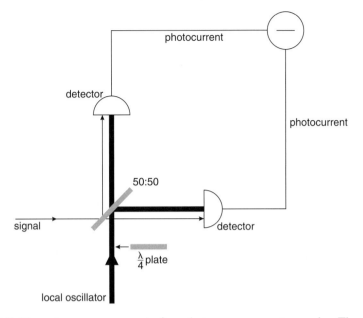

Fig. 3.2 Homodyne measurement of quadrature components x and p. The difference of the photocurrents from the two detectors is proportional to x, and with an inserted quarter-wave plate it is proportional to p.

solely compensate. (Of course, this requires the beamsplitter to be well balanced, which means that the transmittivity and the reflectivity must be equal with high accuracy. This is, in fact, a challenge to experimentalists.) What remains is a current that was theoretically shown to be proportional to the rotated quadrature component x_Θ (see (2.38)) multiplied by the (real) amplitude of the local oscillator, where the angle Θ is just given by the relative phase between the signal and the local oscillator. So this is a wonderful result which enables us to measure *arbitrary* quadrature components in a rather simple way. Once the experimental device is set up, we need only adjust the phase of the local oscillator, which is readily done with the help of a (classical) phase shifter, e.g., by simply displacing a deviating mirror in the path of the local oscillator beam.

It should be noticed, however, that the establishment of a fixed phase between the signal and the local oscillator is by no means a simple task, since a local oscillator with constant phase does not exist. Even the phase of a frequency-stabilized laser undergoes uncontrollable jumps in the course of time. This difficulty can be overcome only in such experimental situations where both the signal and the local oscillator originate from one and the same laser beam. For instance, one part of the latter may be frequency-doubled to serve as a pump in nondegenerate parametric down-conversion used to produce the signal, while the remaining part acts as a local oscillator. In those circumstances, the phase fluctuations of the laser will do no harm, since they are equally transferred to both the signal and the local oscillator, and hence will compensate in the relative phase.

3.3 Utilizing resonance fluorescence

3.3.1 Atomic state detection

The measurement procedures considered thus far have a fatal effect on the object. For photons it is the worst-case scenario; they are completely destroyed in a photodetector, and charged particles detected in a Geiger counter get lost in the electric discharge current they initiated. There are, however, also measurements in which the object survives. These are indirect measurements: the object is coupled to a second system (a so-called meter) on which the measurement is actually carried out. A simple example is the observation of the decay of an excited atomic state through spontaneous emission. Registering the emitted photon with the help of a photodetector gives us the information that the atom has arrived in its ground state. An experimental realization meets, however, serious difficulties. First, an atom emits a photon in an unpredictable direction, and second, the detector has a nonunit detection efficiency. As a result, most of the events will escape the detector's attention.

Fortunately, there exists a very efficient mechanism that allows us to check reliably whether an atom is in a given energy state or not. What I mean is resonance fluorescence induced by a strong coherent field (in practice, laser radiation). Let me first say a few words about resonance fluorescence. It is observed when intense coherent light is interacting resonantly with an atom. This means that the laser frequency is tuned to an atomic dipole transition, say between the energy states $|E_2\rangle$ and $|E_1\rangle$ $(E_2 > E_1)$. Resonance fluorescence is the following process: when the atom is initially in the lower state $|E_1\rangle$, it becomes excited by the strong field to the upper state $|E_2\rangle$, from where it goes back to $|E_1\rangle$, thereby emitting a photon to the side, and this cycle is repeated again and again. Apart from the first step, the same process will happen when the atom starts from the upper level. Now, the point is that the number of photons emitted per second is very large, thus the photons form a *macroscopic* measuring signal. Actually, we can even tolerate that only part of the photons hit the detector's sensitive surface and, moreover, that the detector efficiency is rather low. Notwithstanding those drawbacks, we detect the presence (or onset) of resonance radiation with 100 per cent efficiency!

Two types of measurement can be performed. (*i*) Irradiating an atom with a strong laser pulse, the appearance of a fluorescence signal informs us that the atom is cycling between $|E_1\rangle$ and $|E_2\rangle$, whereas from the absence of such a signal we will conclude that the atom is in any other state, with certainty. When we know beforehand that the atom can only be in, say the state $|E_1\rangle$ or in a state $|\tilde{E}\rangle$ that is different from $|E_1\rangle$ and $|E_2\rangle$ the measurement tells us either that the atom is in the state $|E_1\rangle$ or that it is in the state $|\tilde{E}\rangle$. This kind of measurement is of special relevance in quantum computing (see Section 10.3). (*ii*) When we irradiate the atom continuously, the onset or cessation of fluorescence indicates that the atom has arrived in one of the states $|E_1\rangle, |E_2\rangle$, as a result of whatever interaction, or has made a transition to any other state. This behaviour will be illustrated in the following section.

3.3.2 Quantum jumps

We consider the special situation that resonance fluorescence on a strong transition competes with a weak spontaneous transition that brings the atom from $|E_2\rangle$ to a lower-lying third state $|E_3\rangle$ (see Fig. 3.3a). Thus the atom has a real, though small, chance of escaping the grip of the field that produces the fluorescent light, via the transition $|E_2\rangle \rightarrow |E_3\rangle$. Let us further assume that the atom has an opportunity of returning from $|E_3\rangle$, through additional spontaneous processes, to $|E_1\rangle$. For definiteness, we focus on the simple case that there exists a spontaneous transition to $|E_1\rangle$ which we assume to be very weak; for instance, it may be an electric quadrupole transition. Since we know, especially from radioactive decay, that spontaneous transitions are jump-like events, one will expect that the atom jumps after some time to $|E_3\rangle$ and from here, after a while, to $|E_1\rangle$. As long as it stays in the metastable level $|E_3\rangle$, its interaction with the laser field is 'switched off'. Hence the emission of resonance fluorescence photons will stop for a while, then start again, when the atom has returned to the state $|E_1\rangle$, and so on. This is just what one observes. When the atom is initially in its lower state $|E_1\rangle$, one records a fluorescence signal that remains constant for some time, apart from moderate short-time fluctuations, at a high level. Suddenly this signal falls virtually to zero, and after a while it jumps back to the high level, and so forth (see Fig. 3.4). So an alternation of bright and dark periods (one speaks of intermittent fluorescence) is observed. In particular, the spontaneous transition $|E_2\rangle \rightarrow |E_3\rangle$ is indicated as a jump-like event. On the other hand, the onset of fluorescence signals that the atom has actually reached the lower state $|E_1\rangle$ of the laser induced transition.

It should be emphasized that the observation has to be made on a single system, such as an ion captured in a Paul trap, since the lengths of the bright and the dark periods, as well as the jumping times, are unpredictable. Hence the jumping phenomenon will, in fact, disappear when the observation is simultaneously made on several atoms. From the *average* lengths of the mentioned periods, on the other

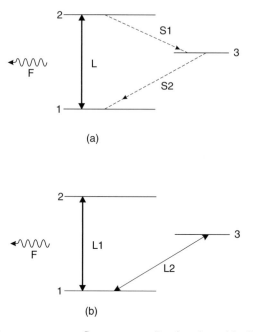

(a)

(b)

Fig. 3.3 Intermittent resonance fluorescence. Depicted are idealized level schemes. (a) L = intense laser radiation that resonantly drives the transition $1 \leftrightarrow 2$;
S1, S2 = weak spontaneous transitions; F = fluorescence light. (b) L1 = intense
laser radiation; L2 = weak laser radiation; F = fluorescence light.

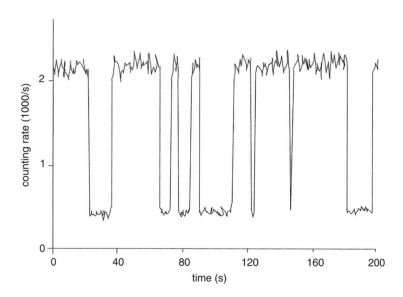

Fig. 3.4 Observed alternation of dark and bright periods in resonance fluorescence
(from Sauter *et al.* (1986)).

hand, one can obviously infer the mean lifetimes, with respect to a spontaneous transition, of $|E_2\rangle$ and $|E_3\rangle$. This is of special relevance for long-lived states, since then the usual determination of the lifetime – and hence the transition probability – from a measurement of the width of the emitted spectral line becomes illusory. In fact, the linewidth is the inverse of the lifetime, and hence unmeasurably small in the present case. For instance, the linewidth corresponding to a lifetime of 1 s is about 1 Hz!

It is important to note that quantum jumps can also be observed when a driven, rather than a spontaneaous, weak transition competes with the strong driven transition $|E_1\rangle \leftrightarrow |E_2\rangle$, giving rise to resonance fluorescence. Let us assume, for instance, that the weak transition, when driven, brings the atom from $|E_1\rangle$ to a higher-lying metastable level $|E_3\rangle$, and from there back again (see Fig. 3.3b). The occurrence of intermittent fluorescence in these conditions may actually be felt as a surprise since it is well known that driving a transition is a *coherent* process (for details see Section 3.3.3). Correspondingly, the atomic wavefunction is a superposition state that in our case can be written as $|\psi\rangle = c_1 |E_1\rangle + c_3 |E_3\rangle$, which means that it is intrinsically uncertain whether the atom is in the state $|E_1\rangle$ or the state $|E_3\rangle$. So this picture is at variance with the observed jumps. What we learn from the experiment is, therefore, that the driven strong transition distinctly disturbs the mentioned coherent evolution, replacing it with jump-like transitions between the two atomic levels taking place at unpredictable instants of time.

I would like to add that the aforementioned type of quantum jump experiment is very helpful when one wants to tune a laser to an extemely narrow atomic line (corresponding, in our scheme, to the transition $|E_1\rangle \leftrightarrow |E_3\rangle$). The problem is how to detect that resonance has actually been achieved. Certainly, the conventional technique of measuring the damping the radiation undergoes as a result of (resonant) absorption cannot be applied. Actually, the desired detection signal is provided by the appearance of dark periods in the resonance fluorescence signal.

3.3.3 Quantum Zeno effect

It was mentioned in Section 3.3.1 that an intense pulse inducing resonance fluorescence acts as a device that measures the energy of an atom. A particular advantage of this kind of measurement is that it can be carried out repeatedly on one and the same system. In the following I will show that a large number of repetitions will affect the atomic behaviour in a strange way.

We will investigate this kind of measurement in some detail. Let us consider again the aforementioned experimental situation depicted in Fig. 3.3b, with the difference, however, that the stationary laser field driving the resonance fluorescence transition $|E_1\rangle \leftrightarrow |E_2\rangle$ is replaced by a short intense pulse, while the weak transition $|E_1\rangle \leftrightarrow |E_3\rangle$ is driven by a long pulse. We assume an ensemble of atoms to be initially in the state $|E_1\rangle$. Somewhat later, the measuring pulse shall be acting on the atoms. There are two possible outcomes of the measurement on a single atom. (*i*) A (pulse-like) resonance signal appears indicating that the atom has been detected in the state $|E_1\rangle$. While the atom cycles between $|E_1\rangle$ and $|E_2\rangle$ as long as the pulse is present, it will fall back to $|E_1\rangle$ after the pulse has passed it. (*ii*) There is no such signal which leads us to the conclusion that the atom is in $|E_3\rangle$. It is noteworthy that this is, in fact, a null measurement. (We postpone a discussion of this amazing kind of measurement to Section 3.4.)

Before the measurement the atoms are in a (coherent) superposition of the states $|E_1\rangle$ and $|E_3\rangle$, and the measuring pulse interrupts the coherent atomic evolution, bringing part of the atoms into $|E_1\rangle$ and the rest into $|E_3\rangle$. Formally, this means that the pure state is transformed into the corresponding mixture, and this 'collapse of the wavefunction' is just what quantum theory declares to be a general feature of the (quantum mechanical) measurement process.

Actually, the measurement in question will hinder the atomic evolution, in the present case the excitation of the atoms. This can be seen from the following argument: as long as the atoms are in a superposition state, they possess coherently oscillating dipole moments[1] (in the sense of the quantum mechanical expectation value, see below), and, just as in classical theory, the work done by the driving field on the induced dipole moments is the energy fed into the atoms. Hence the energy absorbed per second is larger, the larger the dipole moment. The measurement, however, either sets back an individual atom to the initial (lower) state or excites it to the upper state. Both processes result in the destruction of the atomic dipole moment. Thus the atoms must start to build up their dipole moments again, which distinctly reduces the energy transfer compared to the undisturbed evolution. Clearly, this impediment of the atomic evolution will still become enhanced when the measurement is frequently repeated.

This effect is easily studied theoretically, when we treat the field resonantly driving the weak transition $|E_1\rangle \leftrightarrow |E_3\rangle$ as a given classical field remaining unaffected, to a good approximation, by the interaction. Let us first describe the unperturbed evolution. For times t shorter than the pulse duration the wavefunction for the atomic ensemble starting at $t = 0$ from the lower state is given by

$$|\psi_1(t)\rangle = f e^{-i\omega_1 t} |E_1\rangle + g e^{-i\omega_3 t} |E_3\rangle, \tag{3.1}$$

---

[1] When the transition is very weak, it will actually be a multipole, e.g., quadrupole, transition. To simplify the consideration we will focus on a dipole transition.

where

$$f = \cos\left(\frac{\Omega}{2}t\right), \quad g = -\mathrm{i}e^{\mathrm{i}\varphi}\sin\left(\frac{\Omega}{2}t\right). \tag{3.2}$$

Here, $\hbar\omega_1$ and $\hbar\omega_3$ are the energies of the states $|E_1\rangle$ and $|E_3\rangle$, respectively, and φ is the phase of the driving field. The parameter Ω is the Rabi frequency, defined in Section 2.6.4, for the transition in question.

It should be noted that (3.1)–(3.3) also actually apply to a multipole transition. Then the atom couples to the field via the transition matrix element of the multipole moment. What is important is that this coupling is still proportional to the field amplitude A at the atom's position.

The wavefunction for atoms starting from the upper state $|E_3\rangle$, on the other hand, reads

$$|\psi_3(t)\rangle = -g^*e^{-\mathrm{i}\omega_1 t}|E_1\rangle + fe^{-\mathrm{i}\omega_3 t}|E_3\rangle. \tag{3.3}$$

We learn from (3.1) that at time $t = T \equiv \pi/\Omega$ the atoms are in the upper state $|E_3\rangle$ with certainty. So, when a driving pulse has just this duration – it is then called a π pulse (cf. also Section 2.6.4) – it will excite *all* atoms. So this is really the maximum effect a pulse can have on de-excited atoms, and we will later focus on this case. (Note that for much longer pulses the evolution actually becomes reversed: the atoms undergo stimulated emission, which eventually brings them back into the initial state $|E_1\rangle$, from which they start again to run through the cycle, and so forth. It should be noticed, however, that the coherent atomic evolution will be perturbed by atomic relaxation processes, in particular spontaneous emission. Hence it can actually be observed only for times shorter than the relaxation times.)

The induced dipole moment is readily calculated from (3.1) to be

$$\langle\psi_1| ez |\psi_1\rangle = (e^{-\mathrm{i}(\omega_3-\omega_1)t}f^*g + \text{c.c.})D_{13}, \tag{3.4}$$

where the transition matrix element $D_{13} = \langle E_1| ez |E_3\rangle$ has been assumed positive, as in Section 2.6.4. It becomes obvious from this expression that the induced dipole moment vanishes when the atom is either in the lower or the upper level, whereas it becomes maximum when the moduli of f and g are equal, which is the case for $t = T/2$.

Let us now study the effect of measurements carried out with intense short pulses that are resonant with the transition $|E_1\rangle \leftrightarrow |E_2\rangle$. We will assume the pulse duration to be much shorter than T, which allows us to idealize the measurement as a momentary process. A first measurement, at time t, destroys the phase relations in the wavefunction (3.1), transforming it into a mixture $\varrho(t)$ characterized by a diagonal density matrix (with respect to the basis $|E_1\rangle$, $|E_3\rangle$) with matrix elements

$|f|^2$ and $|g|^2$. In fact, the whole ensemble can be decomposed, as a result of the measurement, into two subensembles being in the state $|E_1\rangle$ and $|E_3\rangle$, respectively. What follows is a coherent evolution of the subensembles according to (3.1) and (3.3). A second measurement, τ seconds later, produces a mixture $\varrho(t + \tau)$ that is again described by a diagonal density matrix. The relationship between those two density matrices takes its simplest form when we consider the inversion $I = \rho_{33} - \rho_{11}$, i.e., the difference of the occupation probabilities for the upper and the lower level, respectively. It simply reads

$$I(t + \tau) = \cos(\Omega\tau)I(t). \tag{3.5}$$

We compare now the following two processes:

(*i*) The atoms being initially prepared in the lower state $|E_1\rangle$ so that $I(0) = -1$, are irradiated with a π pulse, without any perturbation by measurement, which brings them all into the upper state $|E_3\rangle$.

(*ii*) The same pulse is applied, but in addition a series of measurements is carried out at times $\tau, 2\tau, \ldots, n\tau = T = \pi/\Omega$. So we are actually dealing with n measurements in neat order, with a delay of $\tau = T/n$ between subsequent measurements. Then we conclude from (3.5) that the inversion at time T, when the π pulse has passed, is given by

$$I(T) = \cos^n(\Omega\tau)I(0) = -\cos^n(\pi/n). \tag{3.6}$$

Now, the point is that $\cos^n(\pi/n)$ goes to unity for $n \to \infty$. (It can actually be approximated by $\exp(-\pi^2/2n)$.) Hence, the occupation probability for the upper level,

$$\rho_{33}(T) = \frac{1}{2}[1 + I(T)] = \frac{1}{2}[1 - \cos^n(\pi/n)], \tag{3.7}$$

decreases with growing n, thereby approaching the initial value 0. So we can say that for large values of n virtually nothing has happened; the motion of the atoms has become 'frozen'.

Similarly, atoms that are initially in the excited state remain excited when subjected to frequently repeated measurements. Amazingly, the observations are mostly null measurements in this case, i.e., the measurement result is indicated by the absence of a fluorescence signal.

The theoretical predictions were, in fact, impressively confirmed experimentally by Itano *et al.* (1990) for different values of n. The measurements were carried out on about 5000 Be ions driven by an rf pulse. Optical pulses were used for measurement. The ions were stored in a Penning trap and laser cooled. In these conditions, their mutual interaction can be neglected, which is a tacit (nevertheless basic) assumption in the above theoretical analysis. (Actually, ten years later the impediment of the coherent atomic dynamics was demonstrated by Balzer *et al.* (2000) on a single trapped Yb ion upon probing it repeatedly.) Itano *et al.* (1990) observed complete inhibition of the motion from level 1 to level 3, within experimental error, for the maximum number of measuring pulses applied, namely $n = 64$.

So freezing of motion through frequently repeated observations is an experimental fact. It reminded some theorists of the paradox of the flying arrow put forward by the Greek philosopher Zeno. Hence they baptized the effect the 'quantum Zeno effect'. It should be emphasized, however, that the flight of an arrow was felt by Zeno as a logical, rather than a physical, paradox, his argument, in essence, being that, since it has a certain position at any instant of time, the arrow is constantly at rest and hence cannot move towards its target. (Actually, it needed differential calculus to overcome this seeming paradox.)

By the way, we also encounter a factor of the type $\cos^n(\alpha/n)$ in classical optics. Consider the following experiment: let a vertically polarized light beam fall on a polarizer that is rotated by an angle α. Then the transmitted amplitude will be damped by the factor $\cos\alpha$. Now, what will happen when we replace the polarizer by a set of n polarizers rotated by $\alpha/n, 2\alpha/n, \ldots, \alpha$? Evidently, the damping factor is now $\cos^n(\alpha/n)$ which tends to unity for $n \to \infty$, as was mentioned above. So we arrive at the surprising result that for sufficiently large n virtually no damping takes place (provided, of course, the polarizers are free from absorption), whereas the effect of just a single polarizer rotated at the full angle α may be quite drastic. In particular, for the special case $\alpha = \pi/2$ no light is transmitted at all.

Finally, it is important to note that an impediment of the quantum mechanical motion through measurements does not occur in spontaneous transitions. In fact, the latter are incoherent. This means that, in contrast to the coherent transitions studied above, no coherently oscillating atomic dipole moments are evolving during the interaction (the quantum mechanical expectation value of the dipole operator vanishes all the time). So there is no impeding mechanism such as the destruction of an existing dipole moment through measurement. (The same holds true, of course, for multipole transitions.)

3.4 Null measurement

We learned in the preceding section that the quantum jump technique gives us a very effective measurement scheme to hand that enables us to investigate in which energy state an individual atom or ion is. The measurement signal is an intense resonance fluorescence pulse, yet also the absence of such a pulse, in the case where the laser pulse driving the resonance fluorescence is switched on, gives us information on the atomic state. We are then dealing with a null measurement. Strictly speaking, the null result tells us only that the atom is *not* in one of the two levels $|E_1\rangle$ or $|E_2\rangle$ (or in a quantum mechanical superposition of them) involved in resonance fluorescence. So it is only when we have a priori information on the atom (of the kind that it has access to just one third level $|E_3\rangle$) that we can interpret the null measurement as a true observation indicating that the atom is in the third level. Amazingly, however, we infer this measurement result from the complete *absence* of a physical signal.

Things become really mysterious when the system is initially in a superposition of the states $|E_1\rangle$ and $|E_3\rangle$. Then we should call to mind that quantum theory associates any measurement with a reduction of the wavepacket. So in the case of a null measurement nothing happens, and nevertheless the wavefunction collapses? What kind of physics is this? However, from the experimental point of view, there is no mystery. You simply cannot observe, *on a single system*, any effect associated with the reduction of the wavepacket. In fact, what happens, according to theory, is that an intrinsically indeterminate variable (in the considered case the energy) becomes sharp. This is no process to be followed by measurement since you cannot detect, on an individual system, that a certain variable is indeterminate. It is only on an ensemble that a transition from indefiniteness to definiteness becomes observable. For instance, the coherent radiation from the macroscopic dipole moment, which is the sum of the individual microscopic dipole moments, ceases abruptly when an energy measurement is carried out. However, observing an ensemble, null results will be observed only on part of the ensemble, the remaining systems being, in fact, affected by the measurement. Hence, apart from the less interesting case that the whole ensemble is in an eigenstate to be detected through a null measurement, the measurement actually disturbs the ensemble. Also reduction of the wavefunction takes place: since some systems undergo a null measurement whereas others give rise to a measurement signal indicating a different outcome, the initial ensemble is transformed into a mixture of two subensembles with different physical properties.

In fact, there are more examples of null measurements. Consider the action of a beamsplitter, say a 50:50 beamsplitter. Loosely speaking, an incoming photon will be either reflected or transmitted. Actually, it is intrinsically uncertain in which of the two output ports the photon leaves the beamsplitter. Otherwise we could not understand that a photon 'interferes with itself'. Accordingly, the ensemble of photons emerging from the beamsplitter is described by a superposition of states corresponding to reflection and transmission, respectively. Placing, now, a detector (assumed to have 100 per cent efficiency) in just one output port, we will register a photon in half of the cases. In the remaining cases a null measurement takes place: the detector does not respond, and from this we infer that the photon must be in the other output port. (To make this conclusion stringent, we must be sure that a photon actually entered the beamsplitter. This can be achieved by starting from photon pairs, as they are produced in spontaneous parametric down-conversion, and detecting one member of the pair, cf. Section 2.6.5.)

Actually, the first to draw attention to null measurements was M. Renninger (1960), who considered the following gedanken experiment: an excited atom that is spontaneously emitting a photon is surrounded by an absorbing screen in the form of a sphere with a hole. Apart from diffraction effects, the hole restricts the possible propagation directions of the photon to a cone. This means a drastic alteration of the photon's wavefunction, which describes a spherical wave in the absence of the screen. So the screen causes a reduction of the wavefunction. Since the photon can be absorbed by the screen as a whole only, in any individual case the photon will either vanish as a result of an absorption process taking place anywhere in the screen, or it will emerge from the hole as a whole. Focusing on the latter event, we are dealing with what Renninger calls a 'negative' observation. Nothing happens with the measuring apparatus – to make the absorbing screen a measuring device, it might be fully covered with photodetector arrays – but nevertheless the physical properties of the photon have changed drastically.

It is interesting to compare the quantum mechanical and the classical description. In classical theory, any spherical wave emitted from a point-like source is affected by the absorbing screen. It will be partly absorbed and partly transmitted. This behaviour is reflected by an *ensemble* of photons. However, when we are dealing with a single photon and find it outside the screen, we can be sure that definitely no absorption has taken place. It is the all-or-nothing principle that quantum theory adds as a novel feature to classical optics.

Finally, it should be noted that measurements on entangled systems bear some resemblance to null measurements. A measurement of a variable of the first subsystem gives us also the value of the corresponding variable of the second (unobserved) subsystem (which is again indeterminate before the measurement) provided we know the entangled state (assumed to be a pure state). So we can get precise information on the second subsystem without acting on it. This opportunity lies, in fact, at the heart of the Einstein–Podolsky–Rosen paradox (see Section 5.2).

3.5 Simultaneous measurement of conjugate variables

According to quantum theory, a pair of variables can be measured simultaneously on the same individual system only when the corresponding Hermitian operators commute. Otherwise, especially when we are dealing with canonically conjugate variables, such as the position and momentum of a particle, that fulfil Heisenberg's commutation relation (see (2.1)), a simultaneous measurement is not possible. This is obvious to the experimentalist, since two measuring devices that are basically different would be needed. In fact, they are mutually exclusive, and hence cannot be applied simultaneously to the same system.

Nevertheless, Arthurs and Kelly (1969), two researchers at Bell Labs., focusing on a simultaneous measurement of position x and momentum p, found a way out of this dilemma, their idea being to make the system under observation S interact with two 'meters' that couple to S. Actually, these are also quantum systems. They can be read out independently and simultaneously (which requires quantum measurements!), thus giving us measured values of the position x_{meas} and the momentum p_{meas}, respectively, of S. (We all know this measuring technique very well from everyday life: to measure the length of an object, we bring it in contact with a ruler and read the latter out.)

The authors chose, in a one-dimensional model, a simple interaction Hamiltonian for the coupling of the meters which allowed for a closed-form solution of the Schrödinger equation. The meters were assumed to be initially in Gaussian states. The most important result of the analysis was the insight that the meters introduce additional noise, i.e., the measured values of position and momentum, x_{meas} and p_{meas}, exhibit stronger fluctuations than x and p. Obviously, this is the price we have to pay for the opportunity of measuring position and momentum simultaneously. The lesson is that we can do it only at the cost of measurement precision. Actually, the authors were able to quantify the extra noise: they found that x_{meas} and p_{meas} obey an uncertainty relation that differs from Heisenberg's in that the right-hand side is enhanced by a factor of 2. So instead of Heisenberg's uncertainty relation

$$\Delta x \Delta p \geq \hbar \frac{1}{2}, \tag{3.8}$$

we have

$$\Delta x_{\text{meas}} \Delta p_{\text{meas}} \geq \hbar. \tag{3.9}$$

It is important to note that the inequality (3.9) holds quite generally, irrespective of what the meters might be and how they are coupled to the system S. The only important thing is that, according to the general rules of quantum mechanics, the actually measured variables x_{meas} and p_{meas} must commute, since they can be simultaneously measured. (Experimentally this becomes possible since the two measurements are executed separately on two different meters.) This requirement already suffices to derive the inequality (3.9), as was also shown by Arthurs and Kelly (1965).

Their argument was as follows. Actually, x_{meas} and p_{meas} are 'fuzzy' variables. So they can be written in the form

$$\hat{x}_{\text{meas}} = \hat{x} + \hat{A}, \quad \hat{p}_{\text{meas}} = \hat{p} + \hat{B}, \tag{3.10}$$

where $\hat{x}$ and $\hat{p}$ are the familiar position and momentum operators and $\hat{A}$ and $\hat{B}$ are 'noise' operators with vanishing expectation values. Starting from (3.10), it needs only some algebra to prove (3.9), whereby (3.8) and the fact that the fluctuations described by $\hat{A}$ and $\hat{B}$ are statistically independent of the system variables x and p, have to be taken into account.

Fortunately, there exists a feasible scheme to measure simultaneously conjugate variables in quantum optics. We start from the idea that such a kind of measurement would be very simple if we were able to clone the original system. Then we could measure one variable on the original system and the other on its clone. Unfortunately, there is a hitch to it: quantum mechanics states that cloning is impossible (cf. Section 5.3). However, this does not exclude the possibility of producing copies that are 'blurred' but nevertheless similar to the original system.

Such a scheme is, in fact, well known in quantum optics. It is the division of an optical beam, with the help of a beamsplitter, into two partial beams. Notwithstanding the fact that these are normally strongly correlated (entangled), they are well separated in space. So nothing hinders you from making separate measurements on them.

The first to realize this scheme experimentally were Walker and Carroll (1984), who determined simultaneously the quadrature components x and p of an optical field that are, in fact, conjugate variables (see Section 2.7.2). They actually measured x on one partial beam and p on the other, and in order to assign those values to the original field they rescaled them, thus undoing the reduction the quadrature components suffered in the splitting process.

The quantum mechanical description of the experiment becomes amazingly simple in the Wigner formalism (Leonhardt and Paul, 1993; cf. also Leonhardt, 1997). Since beamsplitting is a linear process, the evolution of the initial Wigner function is easily described, as explained in Section 2.4. Now, the Wigner function after splitting is a function of both the quadrature components for the first partial beam, x_1 and p_1, and the quadrature components for the second beam, x_2 and p_2. What we have to do is to measure x and p on different beams, say x_1 and p_2. To get the joint probability distribution for those two quadrature components (which describes what we measure on an ensemble), we have to average over the unobserved quadrature components of the two beams (cf. (2.29)). Rewritten for the rescaled measured values, this joint probability distribution turns out to be the initial Wigner function convoluted with a Gaussian. This convolution accounts for the vacuum noise that has entered the unused input port of the beamsplitter, as was explained in Section 2.4. Under ideal experimental conditions (lossless 50:50 beamsplitter, 100 per cent detection efficiency) the smoothing process leads to the Q function (see Section 2.4). This result is just what Arthurs and Kelly (1965) arrived at in their analysis (for balanced meters).

3.6 Characteristic features of quantum measurement

3.6.1 Discrete measured values

In classical physics it is assumed that the variables of any physical system have precisely defined values that may take on, in principle, any real number. That is, the possible values form a continuum. Strictly speaking, a realistic measurement can never yield just one sharp value (one real number); instead it can only indicate that the measured value lies within a certain interval. However, there is no fundamental lower bound for the length of this interval. Improving the measuring technique, you will attain higher measurement precision. In quantum mechanics we face a similar situation in particular cases only, especially when measuring position or momentum of a particle. The measurement of fundamental physical quantities such as atomic or molecular energy and spin, on the contrary, yields discrete values.

To my mind, spin quantization, as described in Section 3.1.1, is the most striking effect. I cannot see any physical reason why angular momentum, supposed to be characterized by a vector of given length, could not be oriented arbitrarily so that its projection onto a chosen direction might take any value between a minimum and a maximum. Quantum theory, however, predicts discrete values as the only possible (measurable) ones from purely formal arguments, namely quantization rules that specify commutation relations for the Cartesian components of the angular momentum. In fact, for the orbital angular momentum of a particle, those commutation relations follow directly from the commutation relations for position and momentum. So it is natural to postulate them also for the intrinsic angular momentum (spin) of elementary particles, and, quite generally, for any kind of total spin that is composed of orbital and intrinsic angular momenta. They are given by

$$\left[\hat{S}_x, \hat{S}_y\right] \equiv \hat{S}_x\hat{S}_y - \hat{S}_y\hat{S}_x = i\hbar\hat{S}_z,$$

$$\left[\hat{S}_y, \hat{S}_z\right] \equiv \hat{S}_y\hat{S}_z - \hat{S}_z\hat{S}_y = i\hbar\hat{S}_x, \tag{3.11}$$

$$\left[\hat{S}_z, \hat{S}_x\right] \equiv \hat{S}_z\hat{S}_x - \hat{S}_x\hat{S}_z = i\hbar\hat{S}_y,$$

where the operator $\hat{S}_x$ describes the x component of the spin vector, etc. Actually, the experiment confirms the spin quantization thus predicted.

From the classical point of view, it appears paradoxical that a measurement on a spin system – for simplicity, let us focus on a spin $\frac{1}{2}$ system such as an electron – will yield $-\frac{1}{2}$ or $+\frac{1}{2}$ (in units of $\hbar$) for the spin component with respect to any direction we might choose. So, interrogating the system with a Stern–Gerlach apparatus (see Section 3.1.1), we may orient the magnet as we like. We will always get either the answer: the spin component has the value $-\frac{1}{2}$ or the answer: the spin component has the value $+\frac{1}{2}$.

As was mentioned in Section 3.1.2, there is a close analogy between spin measurement and polarization measurement. We cannot ask a photon for its polarization. (We simply do not know how to do this.) What we can only do is to ask it: are you x polarized or y polarized? Equally well, we may ask: are you x' polarized or y' polarized? (The prime refers to a rotated coordinate system.) A further possibility is to ask a photon whether it is left-handed or right-handed circularly polarized. But we are granted only one question!

Quantum theory, moreover, tells us that the spin component in a direction that is transverse to that chosen as the projection axis in the experiment is completely undetermined. However, had we decided to orient the Stern–Gerlach magnet such that we measure the mentioned spin component, we would also have found a sharp value, $-\frac{1}{2}$ or $+\frac{1}{2}$.

What has been said about spin or polarization measurements is clearly incompatible with the concept of 'objective reality', which is actually a basic credo in classical physics. What is indicated by those measurements is what has been termed the 'nonobjectifiability' of the properties of microscopic systems.

3.6.2 Back action

A characteristic feature of a quantum measurement is an uncontrollable back action of the measuring apparatus on the observed system. (In realistic cases, such as photoelectric detection of photons and detection of massive particles via impact ionization, this effect is even disastrous.) Even an 'ideal' measurement, as described by the formulae given in Section 2.1, distinctly disturbs the object. This becomes evident from the 'axiom' that a measurement of an observable A brings the object into an eigenstate of the corresponding Hermitian operator $\hat{A}$. Then the conjugate observable B is completely indeterminate, as follows from the uncertainty relation for A and B. This formal arguing, however, tells us nothing about the physical mechanism that produces the uncertainty of B.

Heisenberg discussed this question in the case of position measurement. In a gedanken experiment, he chose a microscope as a measuring device. Actually, making use of Abbe's theory of microscopic imaging, he succeeded in recovering the uncertainty relation for position and momentum of the observed particle. In a microscope, an object in the object plane is imaged onto the image plane through light scattered from the object. So the simplest measuring process is the scattering of just one photon. This means that its propagation direction becomes uncertain, spreading over a solid angle that, in the optimum case, coincides with the collecting angle of the microscopic objective. So the mechanism that disturbs the object becomes obvious: thanks to momentum conservation, the scattering of the photon (which possesses the momentum h/λ, where λ is the wavelength) is associated with a momentum transfer to the particle. As a result, the particle acquires, in particular, an unpredictable amount of transverse momentum. According to Abbe's theory, the resolving power of a microscope (that is, the smallest resolvable lateral distance between two object points) is given by the ratio of λ and the numerical aperture. Hence, to attain a small uncertainty of the position measurement, Δx, one should use light of short wavelength for illumination and make the collecting angle of the lens large. Evidently, both measures serve to enhance the momentum uncertainty, Δp_x, and this is in perfect agreement with Heisenberg's uncertainty relation.

As a second example, let us consider an energy measurement on a quasimonochromatic light field, which we assume to have the form of a square pulse. In what follows, we will call this field the signal. A precise measurement will indicate a certain number N_s of photons to be present. So an ideal measurement leaves the field in an N_s photon (Fock) state. It follows from theory that the expectation value for the electric field strength vanishes for such a state. This means that the phase of the field is completely random. When the original signal is coherent (for instance, it might have been obtained from laser light by pulse shaping and attenuation), its phase has a small uncertainty. So the disturbing effect of the measurement is to destroy the original phase.

This is easy to understand when the measurement (which, anyway, has to be nondestructive) is based on the optical Kerr effect, i.e., a change of the index of refraction of a medium, Δn, induced by the signal. Since Δn is proportional to the light intensity, and hence in the case of a square pulse also to the number of photons contained in the pulse, we are indeed able to measure (indirectly) N_s. To this end, we determine the change of the index of refraction, Δn_p, with the help of a probe wave at a frequency that differs from the signal frequency. (We choose a medium with the property that a field does not induce a change of the index of refraction at its own frequency. So we need not worry about self-interaction effects such as self-phase modulation.) Now, the standard technique to measure the index of refraction is to use a two-path interferometer. Placing the medium in one arm of the interferometer, one produces a phase shift that leads to a shift of the interference pattern. Experimentally, it will be advantageous to use a Mach–Zehnder interferometer. Then the phase shift can be determined from the ratio of the photon numbers in the two output channels.

So an energy measurement on the signal can be carried out in the following way: a wave at the probe frequency is sent onto the entrance mirror of the interferometer (see Fig. 3.5). That part of this wave that travels through the arm containing the nonlinear crystal is our probe wave. Let us assume that the interferometer mirrors M are completely transparent at the signal frequency, whereas they perfectly reflect

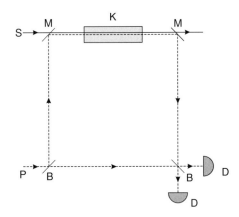

Fig. 3.5 Energy measurement on a light pulse. S = signal; P = probe; B = beam-splitter; M = mirror that acts on the probe only; K = optical Kerr medium; D = detector.

the probe wave. Then we can make the signal copropagate with the probe through the nonlinear crystal. This allows us to measure, through observation of the phase shift, the value of $\Delta n_p = \chi N_s$ (χ is a parameter proportional to the third-order susceptibility of the Kerr medium), and hence N_s. (For more details see Mandel and Wolf (1995).)

So far, so good. However, the probe inevitably acts back on the signal. It induces a change of the index of refraction for the signal, $\Delta n_s = \chi N_p$, where N_p is the number of photons in the probe pulse (assumed to have the same shape as the signal pulse). Now, the point is that N_p cannot be sharp. Even if it would be possible to prepare the wave entering the interferometer in a Fock state, the entrance mirror would make the photon number strongly fluctuate in each interferometer arm. The best we can do is to work with a laser pulse. Then in both arms the photon number fluctuates, to a good approximation, according to a Poissonian distribution (cf. Section 2.2). So we have $\Delta N_p = \sqrt{\bar{N}_p}$, where $\bar{N}_p$ is the mean number of probe photons. As we have seen in Section 1.4, we need a large number of photons to produce a distinct interference pattern. Similarly, in a Mach–Zehnder interferometer the observed photon numbers must be large in order to determine their ratio precisely. But then also ΔN_p will be large. This results in a large fluctuation of Δn_s which, in turn, makes the signal phase uncertain, as predicted by the general quantum mechanical theory of measurement.

3.6.3 Reduction of the wavefunction

The rule that an (ideal) measurement of an observable A brings the system into an eigenstate of the corresponding Hermitian operator $\hat{A}$ was mentioned in Section 2.1

already. Let me comment a little on this rule which, actually, appears as a foreign body in Schrödinger's theory. I would say, *if* a measurement really takes place in the familiar (classical) sense that a (necessarily macroscopic) measuring apparatus indicates unambiguously just one measuring result, the reduction rule will give us the correct description. What is almost miraculous, in my opinion, is, however, that such measuring processes really exist.

Let us consider an instructive example: a beamsplitter acting on an electron produces formally a superposition of two wavepackets that, during their evolution, may become separated in space over a manifestly macroscopic distance. (The latter is unlimited, in principle.) In a suitable (ideal) position measurement the electron will be found at a position that lies *either* within one wavepacket *or* within the other. So the quantum mechanical uncertainty (extending over a macroscopic distance!) is destroyed all of a sudden. From a classical point of view, this can happen only when the electron has definitely moved along one of the possible paths before the measurement. However, this assumption is incompatible with the experimental fact that both wavepackets can be reunited – as is the case in an interferometer – whereby the resulting wavefunction (and hence also the resulting interference pattern) depends on the difference of the phases the wavepackets acquired during their travel. So the physical process described by the reduction of the wavefunction is hard to understand.

In analogy to human behaviour we might characterize the electron's behaviour as follows: the electron cannot 'make up its mind' as to whether it will stay in the first or the second wavepacket. So it leaves the question undecided. However, a suitable macroscopic system with whom it is *able* to interact, forces it to make a decision. Thereby, it is amazing that a detector placed in one of the paths actually suffices to enforce the decision. This means that it affects the electron even when it doesn't respond (provided its detection efficiency reaches 100 per cent). This was pointed out in Section 3.4.

In spite of the headache the reduction postulate might cause, I think we should be happy to have such a tool at hand, which is precisely defined in mathematical terms, and, moreover, easy to handle. Of course, we have to calculate the wavefunction beforehand, and this is normally complicated enough!

I would like to mention that the idea has been put forward that state reduction is ultimately a mental process. It is only when an observer perceives the position of the pointer of the measuring device that the reduction takes place. This concept introduces a subjective element as an indispensable component into the description of Nature. Actually, it is at variance with the modern art of measuring, where the measured data are fed 'online' into a computer for further evaluation. So, fortunately, no human being is needed to take part actively, with his brain, in measurements. Thank goodness, since in view of the overwhelming numbers of data, he would be hopelessly overstrained. So let us take the reduction rule as a wonderful recipe that, in fact, has never failed.

Let me now describe the effect the reduction, or collapse, of the wavefunction has on quantum mechanical ensembles. Formally, the measurement transforms the wavefunction, or the density operator, into a mixture of eigenstates of the measured observable A. This reflects the fact that the systems can be sorted according to the outcomes of the measurement, to form subensembles that are physically discernible. The analogue of a single measurement in classical physics is to select just one subensemble. This is formally described by singling out that eigenstate of A whose eigenvalue coincides with the outcome of the measurement.

It should be emphasized that the destruction of coherence indicated by the vanishing of the nondiagonal matrix elements, in A representation, which is, in fact, the essence of the reduction process, has observable consequences, such as the loss of the ability to interfere (see Section 1.4) or the destruction of an induced (macroscopic) atomic dipole moment (see Section 3.3.3). So the reduction of the wavefunction is indeed a physical process.

An essential point is that the reduction of the wavefunction brings into play the irreversible nature of the quantum mechanical measurement process. Out of all the possibilities that are, so to speak, virtually present in the initial state, just one becomes a fact. This is indicated through a measurement signal which is an event in the macroscopic world. Hence it can be recorded (for instance, on a magnetic tape) and thus be preserved.

The full potential of the reduction process shows up when we consider measurements on entangled systems. Strictly speaking, we suppose the measurement to be carried out on only one subsystem. Let us focus, for instance, on the polarization entangled two photon state (2.36). When, say the second photon is observed to be x polarized, the reduction of the wavefunction means that the latter collapses to the state $|y\rangle_1 |x\rangle_2$, since only this part of the wavefunction (2.36) is compatible with the outcome of the measurement, whereas the other one is in contradiction. So, the first photon has become y polarized, as a result of the measurement on the second photon. This will be so also in the case of realistic measurements in which the second photon is actually destroyed. The first photon, however, still exists, being at our disposal for further experiments.

Apparently, the measurement has exerted a physical action on the first photon. Still more strikingly, this should be a momentary action. This is unbelievable, when we think of experiments in which the two photons are separated, at the instant of measurement, over large distances (several kilometres were actually attained). In those circumstances, it is physically impossible that the measurement on the second photon affects the first one, since it is a fundamental principle in physics that any action can propagate with the velocity of light, at best. (Otherwise the causality principle would be violated, with disastrous consequences.) Must we, in Einstein's words, resort to 'ghost-like fields' as mediators of a mysterious action?

However, there is no *observable* action at all! What really happens is that the polarization of the first photon, which is initially indeterminate, acquires a sharp value. It was pointed out already in Section 3.4 that you cannot observe this 'process', simply because it is impossible to detect uncertainty on an individual system. This has the important consequence that we cannot utilize the reduction process for faster-than-light signal transmission (see also Section 5.3). Hence the causality principle is not affected.

Finally, it should be mentioned that the reduction of the wavefunction is the basic mechanism underlying quantum teleportation. This will be shown in the subsequent section.

3.6.4 Quantum teleportation

Teleportation ('beaming') of a person, instantaneously and over arbitrarily wide distances, is known as a means of easy travelling from science fiction. Is it pure fantasy, or is there a physical foundation to it? Let me shortly explain what quantum theory tells us on this issue. (For more details, see Paul (2004).)

What quantum theory declares as feasible is teleportation of a *quantum state* rather than a physical system. Hence a system must actually be present at the desired destination so that the quantum state can be 'impressed' on it. In practice, this means that a suitable system must be sent beforehand to this place. The possibility of modifying its quantum state in a desired way is provided by choosing it as a member of an entangled pair (see Sections 2.6.5 and 4.3.1). This allows us to carry out a proper measurement in which both the other member and the signal whose state shall be transferred are involved. The reduction of the wavefunction induced by this measurement is, in fact, the mechanism basic to teleportation.

This scheme was ingeniously devised by Bennett *et al.* (1993) and theoretically exemplified on photons. Their aim was to teleport the (unknown!) polarization state of a given photon. As a vehicle, they used a pair of polarization-entangled photons. Following a nice tradition, we will name the two observers dealing with teleportation Alice (sender) and Bob (receiver). One photon of the pair is sent to Alice, and the other to Bob (see Fig. 3.6). Alice initiates the teleportation process by mixing, with the help of a semitransparent mirror, the signal photon and 'her' photon from the entangled pair. Placing polarization-sensitive detectors in the two output channels of the beamsplitter, she is able to make a suitable measurement on the total system 'signal + entangled pair', which effects a reduction of its wavefunction. There are four different possible outcomes of the measurement, and Alice communicates to Bob, via a public channel, the actual measurement result.

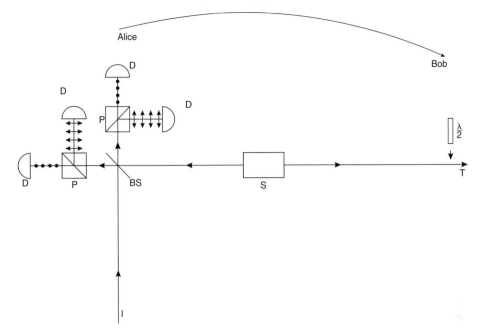

Fig. 3.6 Teleportation of a polarization state. Alice communicates her measurement results to Bob. S = source of entangled photon pairs; BS = beamsplitter; P = polarizing prism; D = detector; I = input signal, T = teleported signal.

Depending on it, Bob has to manipulate 'his' photon, with the help of conventional optical elements such as half-wave plates properly oriented, after instructions given him beforehand. (The latter were actually worked out by a theorist who analyzed the possible results of the reduction of the total wavefunction.) In this way, he brings 'his' photon in the state the signal photon initially had. (Actually, this photon has got lost in the meantime, through Alice's measurement.) Thus teleportation is achieved, at least on paper. Luckily for experimentalists, there is one outcome of Alice's measurement that makes Bob jobless: the photon arriving at his site is already in the signal's state (as a result of Alice's measurement!). This circumstance was exploited in the first experimental verification by Zeilinger's group (Bouwmeester *et al.*, 1997).

Summing up, quantum teleportation exhibits the following characteristic features:

(*i*) Since Alice has to communicate her measurement result to Bob, teleportation needs a finite time not shorter than the transit time of light. So the causality principle is valid.

(*ii*) The scheme works, in particular, when the teleported state is *unknown* to both Alice and Bob.

(*iii*) The signal is destroyed in the teleportation process. In fact, this must be so since otherwise the signal would be cloned, and this would contradict nonobjectifiability which lies at the heart of quantum mechanics (cf. Section 5.3).

Of course, from a practical point of view, one might say, what you achieved through quantum teleportation could have been done much more easily by simply sending the signal photon directly to Bob. The full potential of quantum teleportation will actually show up when dealing with massive particles that travel much slower than light. Then, one will start from an entangled atom pair whose members are sent separately to Alice and Bob beforehand, and the transfer of the quantum state of a signal atom arriving at Alice's site will take place – in idealized conditions – at the velocity of light (since Bob must wait for Alice's message to arrive), whereas a material transport of the signal atom would take much more time.

Quantum teleportation of atomic states was recently realized experimentally by Riebe *et al.* (2004), and, independently, by Barrett *et al.* (2004). The quantum state of the signal atom labelled A – actually, it is an ion – is a superposition of two long-lived energy states $|g\rangle$ and $|e\rangle$, produced through laser irradiation. The teleportation follows the above-mentioned Bennett scheme. The vehicle of transportation is an entangled pair of ions B and C (see Section 4.3.2, (4.9)). All three ions are captured in a radio-frequency trap and cooled down, with the help of laser cooling, to a few μK. They form a linear chain, and hence are vibration coupled. Teleportation of the quantum state of A to C is achieved through a joint energy measurement on A and B. Depending on its outcome, the ion C is in a definite quantum state (as a result of the reduction of the total wavefunction for all three ions), and through a proper manipulation it is brought into the signal ion's original quantum state. (There are four different possible outcomes (gg), (ge), (eg) and (ee), where (ge) means that the ion A is in the ground state $|g\rangle$ and the ion B in the excited state $|e\rangle$, etc., and for each case a definite prescription for how to manipulate the ion C is given beforehand.)

What have been shortly described are certainly pioneering experiments, though one cannot strictly speak of *tele*portation (transport over a wide distance), since the 'teleportation' distance was actually about 10 μm only.

3.6.5 Measurement time

Usually, a theoretician conceives a measurement process to be triggered at an instant that can be chosen at will. The experimenter 'presses a button' to activate the measuring device. Indeed, this is how Wilson's cloud chamber is operated. Here, the activation mechanism is the production of supersaturated steam through adiabatic expansion of the chamber (see Section 1.2.2). Another example is a photodetector that is gated by an external signal.

In many cases, however, the choice of the measurement time is not made by the observer, rather it is left to the system under observation itself. A typical example is the determination of the decay curve for excited atoms with respect to spontaneous emission. Here, a photodetector is placed at a certain distance from the sample, and what is measured is the time between the excitation (e.g., by a short laser pulse) and the first detector click. This click signals that a photon has been absorbed. (The detector is, so to speak, waiting for photons to arrive.) From the click one concludes that (in one atom) a transition from the higher level to a lower one has taken place immediately before. Frequent repetition of this measurement yields a probability distribution for the photon's arrival time, from which the decay law, in particular the mean lifetime of the excited level, can be inferred (cf. Section 1.3).

In the study of the statistical properties of light fields one can get valuable information from measuring the time between two subsequent detector clicks. This is often done with the help of a time-to-amplitude converter (TAC) that works as follows. An electric pulse from the first detector starts the generation of a ramp voltage that grows linearly with time, and a subsequent pulse from the second detector stops this evolution so that the time interval between the two pulses is converted into a voltage height. From the measured intervals one can determine the probability distribution for delayed coincidences, as a function of the delay time.

Finally, it should be emphasized that in quantum mechanics (which, unlike relativistic quantum field theory, is actually a *nonrelativistic* theory!) time is a parameter indicated by a *macroscopic* clock. So, there is a remarkable asymmetry between time and position. Contrary to the latter, time is not quantized. In fact, the introduction of a time operator that is the Hermitian conjugate of the energy operator (for a chosen physical system) meets insurmountable mathematical difficulties. The reason is that both operators differ distinctly in their mathematical structure: while the energy eigenvalues are discrete and have a lower bound (which defines the ground state), the time operator should possess a continuous spectrum of eigenvalues. This discrepancy makes it actually impossible to construct a time operator that fulfils the canonical commutation relation and is, moreover, well behaved. By the way, for a monochromatic radiation field the time operator is, apart from a factor ω (circular frequency of the field), identical to the phase operator. No wonder that all efforts to find an exact phase operator have failed.

4

Correlations

4.1 Field correlations

4.1.1 Optical coherence

When we investigate statistical phenomena, we get valuable information from the study (both theoretical and experimental) of correlations between two, or more, variables. For instance, in classical optics the theory of optical coherence is based solely on the concept of correlations that exist between the electric field strengths oscillating at two positions, r_1 and r_2, in general with a certain delay τ. For stationary fields this correlation is described by the first-order (with respect to intensity) correlation function

$$G^{(1)}(r_1, r_2; \tau) = \overline{E^{(-)}(r_1, t) E^{(+)}(r_2, t + \tau)}. \tag{4.1}$$

The bar denotes time averaging, and $E^{(-)}$, $E^{(+)}$ are the negative- and the positive-frequency parts of the electric field strength $E(r, t)$,

$$E^{(+)}(r, t) = \int_0^\infty f(v; r) e^{-2\pi i v t} dv, \quad E^{(-)}(r, t) = \int_0^\infty f^*(v; r) e^{2\pi i v t} dv = E^{(+)*}(r, t), \tag{4.2}$$

where $f(v; r)$ is the Fourier transform of $E(r, t)$. For simplicity, we have assumed the field to be linearly polarized. The modulus of $G^{(1)}$ has its maximum at $r_2 = r_1, \tau = 0$, from where it falls gradually to zero for $r \equiv |r_2 - r_1|$ or $|\tau|$ going to infinity.

The correlation function normalized to unity at $r_2 = r_1, \tau = 0$, is closely related to interference experiments. Its modulus gives us the visibility of the interference pattern. Of special interest are the following two cases: (*i*) $r_2 \neq r_1, \tau = 0$; then we may think of a Young-type interference experiment with r_1 and r_2 being the positions of the pinholes in the interference screen. (*ii*) $r_2 = r_1, \tau \neq 0$; we face such a situation in a two-path interferometer, for instance a Michelson or a Mach–Zehnder interferometer, with $c\tau$ (c velocity of light) being the difference of the arm lengths. The effective widths of $\left|G^{(1)}(r_1,r_2;0)\right|$, with r_1 and r_2 lying in a plane that is transverse to the propagation direction, and $\left|G^{(1)}(r,r;\tau)\right|$ are the transverse coherence length and the coherence time, respectively.

This is all well known from classical optics. The quantum mechanical description of first-order correlations (in which $E^{(-)}$, $E^{(+)}$ become Hermitian conjugate operators that are normally ordered in (4.1), the time average being replaced by the quantum mechanical expectation value) offers no additional physical insight.

4.1.2 Intensity correlations

Unexpectedly, a fresh impetus to develop quantum optics came from radio astronomy. The British astronomers Brown and Twiss (1956) had the ingenious idea of extending a radio-astronomical observation technique they had developed a few years before, to the optical domain. While their aim was to improve the measurement of the angular diameters of distant stars, quantum theorists got excited about what they soon baptized the Brown–Twiss effect. The basic technique for the determination of the angular diameters of fixed stars had been devised and successfully applied by Michelson long before. It amounts to measuring (on Earth) the transverse coherence length for the light coming from the star. This length is related, in a simple form, to the star's angular diameter. The transverse coherence length is determined from the visibility of the interference pattern that is produced as follows. Two light beams from the star are selected by two mirrors separated by a variable distance d, and made to interfere with the help of deviating mirrors that bring them to the back focal plane of a telescope. Making the distance d larger and larger, one finds a critical distance, d_{crit}, at which the interference pattern virtually disappears. This critical distance is nothing but the transverse coherence length. In fact, this scheme worked very well.

There are, however, severe limitations. When the star's angular diameter is so small that the transverse coherence length is of the order of tens or hundreds of metres, a mechanical stability of the device would be required that is simply utopian. Moreover, the atmospheric scintillations (resulting from local motions of the air) now give rise to a fluctuating optical path difference, which drastically hampers the observation of the interference fringes.

Both problems were solved by Brown and Twiss with a stroke of genius. They replaced the two mirrors by reflectors that focused the star's light onto a photomultiplier each (see Fig. 4.1). On multiplication of the photocurrents and time averaging of the product they obtained their measuring signal. So, what they observed instead of field correlations were, in fact, intensity correlations, the

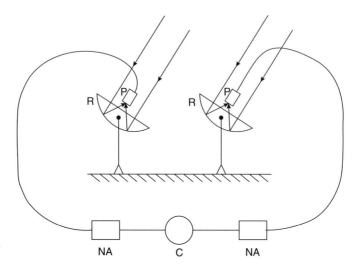

Fig. 4.1 Brown–Twiss stellar interferometer (R = reflector; P = photomultiplier;
C = correlator; NA = narrow-band amplifier). In the correlator the product of the
two photocurrents is taken and time averaged.

equivalent of the disappearance of the interference pattern being now a distinct
decline of the intensity correlations. Clearly, this novel technique eliminated the
effect of atmospheric scintillations, since they affect only the phase of the field and
hence have no influence on the intensity. Moreover, no mechanical construction
ensuring accurately constant path lengths from the mirrors to the interference
plane is needed any longer, the electric signals from the detectors being carried by
cables.

But why should the device used by Brown and Twiss work at all? The reason is
that in thermal, more generally, chaotic, light (in contrast to laser light) phase and
intensity fluctuations go hand in hand. More precisely speaking, over a coherence
volume of the field, defined as a (right-angle) cylinder with the transverse coherence
length as diameter and the longitudinal coherence length (the product of the
coherence time and the velocity of light) as height, the instantaneous (random)
values of *both* the phase and the amplitude, and hence also of the intensity, do
not vary appreciably. The physical background is that the field is produced by
a large number of *independently* radiating emitters (atoms or molecules). In a
classical picture, they oscillate like harmonic oscillators, however, with random
phases. Those phases remain fixed during the emission process. Normally, the
atoms (molecules) emit at different central frequencies (in a gas, this so-called
inhomogeneous line broadening is due to the Doppler effect), so that the phase
relations between the emitted 'elementary' waves are approximately preserved

over time intervals of the order of the inverse bandwidth, i.e., of the coherence time. Hence, during such a time interval the electric field strength resulting from the superposition of the 'elementary' waves varies only a little bit. This means that *both* its phase and amplitude remain virtually constant. It is important to note that the actual value of the amplitude (and hence also the intensity) strongly depends on the concrete values of the (relative) phases between the 'elementary' waves. When the latter interfere constructively, a large amplitude will result, whereas in the case of destructive interference the amplitude will be small. This explains, since the phases of the individual dipole oscillations undergo uncontrollable (jump-like) changes in the course of time, why chaotic light exhibits strong intensity fluctuations.

The intimate connection between phase and intensity correlations can be given a precise mathematical form: in the case of chaotic light, the second-order correlation function

$$G^{(2)}(r_1, r_2; \tau) = \left\langle E^{(-)}(r_1, t) E^{(-)}(r_2, t + \tau) E^{(+)}(r_2, t + \tau) E^{(+)}(r_1, t) \right\rangle, \quad (4.3)$$

which describes the intensity correlations, is related to the first-order correlation function in the form

$$G^{(2)}(r_1, r_2; \tau) = G^{(1)}(r_1, r_1; 0) G^{(1)}(r_2, r_2; 0) + \left| G^{(1)}(r_1, r_2; \tau) \right|^2, \quad (4.4)$$

where $G^{(1)}(r, r; 0)$ is the mean intensity at position r. We have now definitely chosen the quantum mechanical description since, unlike the classical one, it takes properly into account the fact that a detector needs a full energy quantum $h\nu$ to respond. We will soon come back to this point.

Apart from a constant factor, the second-order correlation function, for $\tau = 0$, is what Brown and Twiss observed. It becomes obvious from (4.4) that the decrease of the visibility of the interference pattern is paralleled by a decrease of the intensity correlations. They drop, in fact, to a constant level that is characteristic of random fluctuations. Thus the observation technique introduced by Brown and Twiss is fully substantiated.

It should be noted that (4.4) is a special case of a general relationship saying that, for chaotic light, any correlation function of arbitrarily high order can be expressed through first-order correlation functions. Hence the full information on the statistical properties of the field is already contained in the first-order correlation function.

It soon turned out that the precision of the Brown–Twiss technique could be improved by registering, instead of correlations between photomultiplier currents, coincidence counts of two detectors placed at different positions. This concept is readily extended to include a delay between the two detector clicks. What is thus observed over a long period is the delayed coincidence counting rate.

Above all, delayed coincidences were measured on light fields produced in the laboratory. Of special interest is the counting rate at the same position ($r_2 = r_1$). The problem of placing two detectors at the same position was elegantly solved by Brown and Twiss by dividing the incoming field, with the help of a semitransparent mirror, into two beams sent to separate detectors. The coincidence counting rate thus observed exhibits a distinct maximum at $\tau = 0$ which indicates that the photons show a kind of social behaviour, having a tendency to arrive in pairs. This effect, now known as photon bunching, proved, in fact, to be a great stimulus to quantum theorists. It must be emphasized, however, that photon bunching is not a general feature of light, rather it is characteristic of chaotic, in particular, thermal, light. In fact, the opposite effect, named antibunching, can also be observed under special conditions.

This brings us back to the observation that only the quantum mechanical description of intensity correlations is in agreement with the quantum character of the primary process taking place in a photodetector. This becomes obvious, in particular, when the field is of such a kind that one photon after the other, with a delay that exceeds the response time of the detectors, arrives at the semitransparent mirror in the above-mentioned set-up. Then one will find no coincidences for $\tau = 0$; however, for larger delay times some will be observed. This effect, called photon antibunching, is correctly described by quantum theory (the intensity correlation function $G^{(2)}(r, r; 0)$ vanishes exactly in those conditions). The classical description, on the other hand, fails in those circumstances; it predicts, for *any* stationary field, a delayed coincidence counting rate that has an absolute maximum at $\tau = 0$. Hence the antibunching effect reveals a nonclassical feature of light. Accordingly, light exhibiting this effect falls within the category of what has been termed *nonclassical light*. Experimentally, the antibunching effect is most convincingly demonstrated by investigating resonance fluorescence from a single trapped ion, the first to do this being Diedrich and Walther (1987). In fact, since the ion needs some time to get excited anew, photons are never emitted simultaneously.

A distinct difference between the classical and the quantum mechanical description becomes obvious also when two photons arrive simultaneously (experimentally, this means within a time window given by the detectors' response time) at a beamsplitter (see Fig. 4.2). The amazing result of the quantum mechanical analysis is that both photons are found in either the first or the second exit port (thus forming what has been called a 'biphoton'), but it never happens that each port contains one photon. This was also verified experimentally (Hong, Ou and Mandel, 1987). According to classical theory, however, the latter event is as likely as the creation of a 'biphoton'.

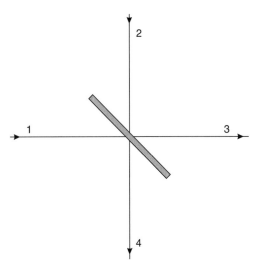

Fig. 4.2 Beamsplitter used as an optical mixer. 1 and 2 are the incoming beams; 3 and 4 are the outgoing beams.

4.1.3 Higher-order interference

It is a well-known fact in classical optics that no interference takes place when the transverse coherence length or the coherence time is exceeded. The term 'interference' is used here as a synonym for the appearance of an interference pattern. Actually, this historical interference concept turned out to be too restricted in view of the modern observation techniques provided by photodetectors. It seems appropriate to speak of interference whenever individual waves become superposed. Let us consider, as a typical example, a Young-type experiment in which the pinholes in the interference screen are separated by more than the transverse coherence length. In such conditions no interference pattern can be observed. (One might say, this is so by definition of the transverse coherence length.)

However, this does not mean the absence of interference in a general sense. In fact, during any short time interval of the order of the coherence time, a definite phase relation will exist between the 'virtual' light sources represented by the illuminated pinholes. Hence, a 'momentary' interference pattern will be produced. However, the relative phase will randomly change in the course of time, and this gives rise to unpredictable shifts of the individual interference patterns. Thus no interference pattern can be discerned when the observation extends over long times (compared with the coherence time). What one actually sees is a superposition of many individual interference patterns that are randomly shifted. The question arises, can we get information on the constituting elements of this superposition?

What offers itself first is to drastically reduce the exposition time, with the help of a shutter, so that a 'momentary' interference pattern becomes observable. This procedure works, however, only at very high intensities. In fact, during the exposition time many photons must be registered in order to detect an interference structure. So what can we do at normal, or even rather low, intensities?

The answer is: observe intensity correlations! The experimental scheme should be as follows. Coincidences are counted between two detectors that are aligned in the interference plane in the direction orthogonal to the 'virtual' interference fringes, that is, the fringes that would appear when the pinholes were illuminated with coherent light. Those coincidences have to be observed for different separations d of the detectors. One will expect that at $d = n\Lambda$ ($n = 1, 2, \ldots$), where Λ is the fringe spacing in the mentioned 'virtual' interference pattern, the coincidence counting rate has a maximum from where it drops, with varying d, to a minimum at $d = (n + \frac{1}{2})\Lambda$. A simple argument is that for $d = n\Lambda$ the instantaneous intensity is the same for both detectors; in particular, there will occur events in which both detectors feel the maximum intensity corresponding to a fringe maximum, and hence have a large probability to respond jointly. For $d = (n + \frac{1}{2})\Lambda$, on the other hand, such lucky events can never happen. On the contrary, when the intensity reaches a maximum at one detector, it will vanish at the other, so that coincidences are impossible in this case.

For a more quantitative estimate of the intensity correlation function, in dependence of d, we make use of classical theory. What we have to do is to multiply the (instantaneous) intensities at the detector positions z_1 and z_2 and average over the relative phase of the two waves that emerge from the two holes and get superposed in the interference plane. (We denote the direction orthogonal to the 'virtual' fringes as the z direction.) A 'momentary' interference pattern is described by the intensity distribution

$$I(z) = I_0[1 + \cos(2\pi z/\Lambda + \Delta\varphi)]. \tag{4.5}$$

On multiplication of $I(z_1)$ and $I(z_2)$ and averaging over $\Delta\varphi$, we readily get the result

$$\overline{I(z_1)I(z_2)} = I_0^2\left[1 + \frac{1}{2}\cos(2\pi(z_2 - z_1)/\Lambda)\right], \tag{4.6}$$

which, in particular, confirms our simple argument.

It is interesting to note that the intensity correlation function (4.6), unlike the intensity distribution (4.5), does not drop to zero, but only to half its average value. A quantum mechanical analysis, however, shows that this is true only for large intensities, the minimum values becoming actually smaller and smaller with decreasing intensity. In the extreme case, when no more than two photons are present during the response time of the detectors, the coincidence counting rate goes down to zero for $d = (n + \frac{1}{2})\Lambda$. This was verified experimentally by Ghosh and Mandel (1987), who worked with photon pairs from parametric down-conversion. The two photons of a pair, emitted in slightly different directions, were brought together with the help of two deviating mirrors. So the quantum character of the electromagnetic field makes the variation of the intensity correlations with d more pronounced, compared with the classical description.

The reason for the failure of the classical theory is that it utilizes a detector concept that is incorrect at very low intensities. It is assumed that the probability of a photodetector to respond is proportional to the intensity, irrespective of the intensity level. Hence there exists a nonvanishing probability to register a coincidence count even when fewer than two photons are present. The quantum mechanical theory, on the contrary, correctly predicts that no coincidence count can be observed in this case, simply because both detectors must absorb a photon each.

4.2 γ–γ angular correlations

While the correlations discussed before yield information on field properties, information on the emission process can be extracted from angular correlations between successively emitted quanta. This technique found application in nuclear physics, where the emission of two γ quanta from an excited nucleus is investigated. Actually, it is a cascade process: the first γ quantum is emitted in a transition that ends in an intermediate level, and the second one starts from this level. Coincidence counts are detected with the help of two separate detectors that are placed at some distance from the probe (see Fig. 4.3). It turns out that the coincidence counting rate varies with the angle θ between the emission directions of the recorded quanta. In the absence of external electric or magnetic fields the nuclear ensemble is not oriented. Hence the coincidence counting rate depends only on the modulus of θ, actually it is a function of $\cos\theta$, from whose specific form the spins and parities of the nuclear levels involved can be inferred.

The measurement scheme relies on the fact that in a spontaneous transition from a level with spin J_2 (in units of $\hbar$) to a lower level with spin J_1 angular momentum is conserved. This means that the total spin in the final state, i.e., the (vectorial) sum of the nuclear spin J_1 and the angular momentum of the emitted quantum L, equals the nuclear spin in the initial state, J_2. As a result of spin quantization, the quantum numbers J_2, J_1 are positive integers (including zero) or half-integers, and the spin of the emitted quantum can take the values $L = 1, 2, \ldots$ corresponding to dipole, quadrupole, ... radiation. The angular momentum conservation law restricts the possible values of L to $L = |J_2 - J_1|, |J_2 - J_1| + 1, \ldots J_2 + J_1$. (Obviously, this restriction requires both J_1 and J_2 to be either both integers or both half-integers.)

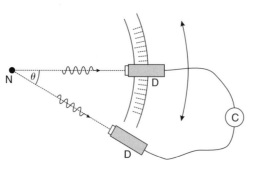

Fig. 4.3 Scheme for the measurement of $\gamma-\gamma$ angular correlations. N = excited
nucleus; D = detector; C = coincidence counter.

In addition, the spin components with respect to the quantization axis, i.e., the
magnetic quantum numbers m_2, m_1 and m corresponding to J_2, J_1 and L, are required
to obey the relation $m = m_2 - m_1$, when the transition starts from sublevel m_2 and
ends in sublevel m_1. So the emitted radiation has multipole character and, hence,
exhibits a specific angular characteristic in a transition between any two sublevels.
Nevertheless, the total radiation emitted in all possible transitions between sublevels
is isotropic, provided that in the initial state all sublevels are equally populated. This
is so, because there is no preferred direction. Note that the quantization axis can be
arbitrarily chosen!

So one will ask, how do the angular correlations in question come about? The
answer is that the observation of the first γ quantum is a selection process: from
the ensemble of all nuclei initially excited, a subensemble is selected for further
observation. For this subensemble, however, the population of the sublevels of the
intermediate state is no longer uniform. As a result, the (total) radiation emitted
from this state will no longer be isotropic, and this is what one observes.

4.3 Entanglement

4.3.1 Natural correlations

Two systems 1 and 2 are said to be entangled when they are in a quantum state
that cannot be factorized into two states (pure or mixed) describing the systems 1
and 2, respectively. (In fact, factorizability means statistical independence.) In this
sense, entanglement is the most natural thing in the (quantum) world, since *any*
interaction between two systems brings them in a nonfactorizable state.

Consider, for example, spontaneous emission from a localized atom. With the
help of a photodetector array surrounding the atom, we can measure the propagation
direction of the emitted photon. (Actually, the detectors 'force' the field to 'make a
decision' on the emission direction.) Together with the frequency assumed known,

this gives us the photon's momentum vector, $p = \hbar k$, where k is the wave vector. From momentum conservation we conclude that the atom has suffered a recoil that has the same magnitude as the photonic momentum, but is oppositely directed. This does not mean, however, that the atomic and the photonic momentum are correlated so strongly that the measurement of one of those quantities allows us to predict *uniquely* the actual value of the other. This is because of the uncertainty of the atomic momentum in the initial state, which is, in fact, unavoidable according to Heisenberg's uncertainty relation, when the atom is well localized. As a result of the recoil, the atomic momentum distribution is shifted, but the uncertainty remains, of course.

It is interesting to note that the impossibility of determining precisely the photon's propagation direction from a measurement on the atom was used to refute a shrewd argument against the incompatibility of which-way information and interference (Pauli, 1933). The argument is, in fact, a modified version of the criticism of quantum mechanics, put forward by Einstein in the famous 1929 Bohr–Einstein debate (Bohr, 1949). We consider a Young-type experiment with a spontaneously emitting atom as an illuminating source. Of course, we have to repeat the experiment many times to get an interference pattern. Whenever a photon has been detected on the interference screen, we might – this is the argument – measure the component, with respect to the direction x defined by the two pinholes in the observation screen, of the atomic recoil. From this we could infer the emission direction of the photon, which would allow us to tell which of the two pinholes in the interference screen the photon actually passed through. (Note that the measurement on the atom in no way affects the photon, which has actually disappeared already.) This argument, however, fails, since it rests on two conflicting assumptions. (*i*) The atom is well localized, in particular, in the x direction, since otherwise the interference pattern will be washed out. (*ii*) The initial uncertainty of the atomic momentum p_x is sufficiently small so that we can reliably determine the recoil, in the x direction, from a measurement of the atomic momentum some time after the emission. A simple analysis reveals that the two assumptions cannot be fulfilled simultaneously, since this would contradict Heisenberg's uncertainty relation.

As a matter of fact, there are composed systems in which two variables u and v belonging to different subsystems are so strongly correlated that a sharp value of u is uniquely associated with a sharp value of v. A typical example is photon pairs produced in spontaneous parametric down-conversion (see Section 2.6.5). It follows from the conservation laws (2.34) and (2.35) that both the energies and the momenta of the two photons fulfil the aforementioned requirement, provided the pump is a monochromatic plane wave, to a good approximation.

It was pointed out in Section 2.6.5 that in the case of type II interaction, the polarization states of the two photons propagating in properly selected directions are entangled too. Actually, the state (2.36) deserves to be called maximally entangled, since the two terms corresponding to different outcomes of a polarization measurement are contributing with equal weights. (If one were dominating, the state would come close to a factorizable, and hence nonentangled, state.)

While the polarization entanglement in question has its origin in the anisotropy of the nonlinear crystal, we also know physical systems where it results from angular momentum conservation. This is so in a $J = 0 \rightarrow J = 1 \rightarrow J = 0$ two-photon cascade (J atomic spin in units of $\hbar$) in which two photons are successively emitted from an excited atom. We may then argue as follows. In the initial state of the cascade no photons are present, and hence the total angular momentum (the sum of the spins of the atom and the electromagnetic field) vanishes. Angular momentum conservation requires that this is so also in the final state. Since J equals zero again, this means that the spin of the emitted two-photon field must be zero.

This must also hold true, in particular, for the spin projection onto a straight line that is defined experimentally by two detectors placed such that the atomic source is amidst them. Obviously, the condition in question is fulfilled when one photon is right-handed circularly polarized (spin component $m = 1$) and the other one left-handed ($m = -1$). (We refer the rotational sense to the same direction.) However, it is indeterminate which of the two photons possesses right-handed circular polarization, and which left-handed. So we are dealing with a (maximally) entangled state of the form

$$|\psi\rangle = \frac{1}{\sqrt{2}}(|+\rangle_1 |-\rangle_2 + |-\rangle_1 |+\rangle_2), \tag{4.7}$$

where $|+\rangle_1$ describes a right-handed circularly polarized photon in beam 1, etc. Since a circularly polarized wave can be considered as a superposition of a wave linearly polarized in one direction (x) and a wave linearly polarized in the orthogonal direction (y), we can readily rewrite the state (4.7) as

$$|\psi\rangle = \frac{1}{\sqrt{2}}(|x\rangle_1 |x\rangle_2 + |y\rangle_1 |y\rangle_2). \tag{4.8}$$

It should be noticed, however, that the emission directions of the two photons are not specified. So the detectors actually select specific photon pairs from a huge manifold of pairs whose members are emitted in unpredictable directions. Fortunately, the situation changes for the better when we use spontaneous parametric down-conversion for photon pair generation (see Section 2.6.5). Then, the emission directions of the two photons are strongly correlated so that, in principle, the direction of only one photon must be preselected with the help of an aperture. Hence this technique is superior to that based on a two-photon cascade. Moreover, also the quantum state (4.8) can be produced in down-conversion.

In fact, it is readily obtained from (2.36) when α is made equal to zero and one of the beams is sent through a half-wave plate that rotates the polarization direction by 90°.

Polarization entanglement exists also in positronium annihilation. Here, an electron–positron pair bound in an s-state (spin zero) is spontaneously converted into two γ quanta. With respect to their polarization properties, the same argument applies as for the two-photon cascade. Owing to momentum conservation, the two quanta propagate in opposite directions. This advantage is, however, overcompensated by the lack of efficient polarizers.

Coming back to the entanglement stemming from energy and momentum conservation, one might be tempted to ask what is really so exciting about the strong correlations of the two photons produced in spontaneous parametric down-conversion. In fact, we observe similar events in classical mechanics! Think of a bomb assumed to burst into two fragments. Thanks to momentum conservation, an observer detecting one fragment knows immediately the direction in which the other fragment is propagating (provided he has a priori information that the bomb was initially at rest and has disintegrated in just two parts). He readily calculates from the observed mass and velocity of 'his' fragment its momentum. Reversing its direction, he gets the momentum of the remaining fragment. When he knows also the mass of the bomb, he knows the mass of the unobserved fragment, and so he can precisely find out its motion in the past as well as predict its future motion. However, there is a fundamental difference between the quantum mechanical and the classical situation. In the latter, the motions of the two fragments will be determinate, in the sense of objective reality. So what the observer does is simply to take notice of already existing facts.

This, however, is not so in the quantum mechanical case. The propagation directions of the two photons are intrinsically uncertain (they are only restricted to cones corresponding to different frequencies), and they take sharp values only as a result of a measurement. Actually, the measured values obey the conservation laws. Hence it suffices to measure the propagation direction of just one photon. As in the classical case, this immediately gives us precise information on the propagation direction of the unobserved photon.

Quantum theory predicts that entanglement is preserved even when the subsystems become separated over large distances during their temporal evolution. Consider, for example, polarization-entangled photon pairs. Coupling the photons in a suitable fibre each, their distance can be made as large as kilometres. It is hard to believe that the quantum mechanical correlations should still exist in such extreme conditions. However, they were actually measured! I think there is a simple argument in favour of the quantum mechanical prediction. As has already been mentioned, polarization entanglement produced in a two-photon cascade

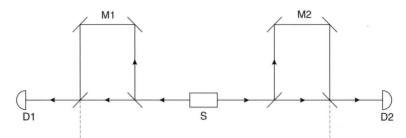

Fig. 4.4 Franson experiment. Coincidences between detectors D1 and D2 are registered. S = source of entangled photon pairs; M1, M2 = identical Mach–Zehnder interferometers.

results from angular momentum conservation. Hence something like a 'spontaneous breakdown' of the polarization correlations at a certain critical distance between the photons would mean no less than that one of the fundamental conservation laws in physics would suddenly be suspended, and this would be outright shocking.

Photon pairs produced in spontaneous parametric down-conversion have still another nonclassical property. Since they are generated in one and the same elementary process, they are emitted simultaneously. However, the instant of emission itself is indeterminate. Thus they are suited for a novel type of experiment. Really amazing is an experiment proposed by Franson (1989) and carried out by Kwiat, Steinberg and Chiao (1993). The two photons belonging to a pair are sent through a Mach–Zehnder interferometer each before they are separately detected, whereby the detectors are placed at the same distance from the source (see Fig. 4.4). Coincidence counts are registered. An interferometer offers an incident photon a short and a long path, their difference giving rise to a transit time difference ΔT.

In practice, the signal and idler photons have large bandwidths so that they are short pulses, in the wave picture. The interferometer dimensions can be chosen such that the pulse durations are short compared with ΔT. Hence the two pulses into which an incident pulse is split at the entrance mirror of the interferometer will never meet again, so that conventional interference (in the sense of the 'interference of the photon with itself') is impossible. However, there is another kind of interference that results from the indeterminacy of the emission time. When a coincidence count is registered at time t, it is intrinsically uncertain whether this event has originated from an emission process at time $t - t_1$, with t_1 being the time taken by a photon propagating from the source to the detector via the short interferometer path, or at time $t - t_1 - \Delta T$.

In a simplified picture we may introduce a probability amplitude $\psi(t)$ for a coincidence count at t, corresponding to the first possibility. The probability amplitude for the case that both photons started ΔT seconds earlier and took the long paths in the interferometers, will have the same modulus; however, it will take notice of the extra phases the photonic electric field strengths accumulated during their passage through the respective resonator. Since the squared modulus of $\psi(t)$ has the meaning of an intensity correlation, $\psi(t)$ will correspond to the product $E_s^{(-)}E_i^{(-)}$ of the negative (or, equivalently, positive) frequency parts of the *classical* electric field strengths of the two photons, the signal and the idler photon (cf. (4.3)). Hence the phase difference, with respect to the photons propagating along the short paths, will be $\Delta\varphi = (\omega_s + \omega_i)\Delta T$. Apparently, $\Delta\varphi$ is a strongly fluctuating quantity since both the (circular) frequencies ω_s and ω_i possess a large uncertainty given by the respective bandwidth. However, the specific character of the emission process subjects the individual frequencies to a drastic restriction: their sum must equal the pump frequency (see (2.34)) which is sharp. This lucky circumstance gives the phase difference a sharp value too, $\Delta\varphi = \omega_p\Delta T$. Needless to say, this 'trick' cannot be reproduced in a classical description.

The rest is simple. In accordance with the linearity of the Schrödinger equation, which allows for superpositions, we have to add the two probability amplitudes to obtain $\psi_{tot}(t) = \psi(t)(1 + e^{i\omega_p\Delta t})$ which gives us, up to a constant factor, the probability itself as $|\psi_{tot}(t)|^2 = 2^{-1}[1 + \cos(\omega_p\Delta T)]$. So we find a simple dependence on ΔT which reminds us of a conventional interference pattern. This result is really amazing: though we are counting photons, we can observe interference fringes when varying ΔT. Experimentally, it is easier to insert in one interferometer (or both) a phase shifter and study the coincidence counting rate in dependence of the additional phase shift (which simply adds to $\omega_p\Delta T$).

4.3.2 Man-made entanglement

From what has been said before one might get the impression that we must leave it to Nature to provide us with entangled systems, the experimenter's task being reduced to provide the conditions in which Nature can act. Fortunately, this is not the last word.

In quantum optics, entangled states can be produced with the help of an amazingly simple instrument, namely a beamsplitter. While it looks rather harmless from the classical viewpoint, it shows amazing capacities in the quantum world. Let us make a single-mode optical field impinge on a beamsplitter assumed lossless. The first astonishing thing is that in the formalism we must explicitly take into account that vacuum couples to the unused input port. This was mentioned already in Section 2.4. As a result, the quantum state for both the reflected and the transmitted field is not factorizable, the only exception being Glauber states. (A Glauber state for the incident field is transformed into the product of two Glauber states, in perfect analogy to classical optics.) The entanglement thus produced becomes obvious when we focus on the measurement of the photon number. When a fixed number of photons n is falling on the beamsplitter (fortunately, the case $n = 2$ is realizable experimentally utilizing degenerate parametric down-conversion), the energy conservation law requires that when you measure k $(= 0, 1, 2, \ldots, n)$ photons in one beam, the number of photons in the other beam must be $n - k$, with certainty.

It is more difficult to manipulate two independent material systems such as to bring them into an entangled state. Only recently, such an experiment was successfully performed with ions stored in a Paul trap (Roos *et al.*, 2004). The latter can be given such a shape that the ions become aligned (linear Paul trap). They will undergo oscillations around their equilibrium positions. Since they are coupled through the repulsive Coulomb force, they will execute collective motions that are described as an excitation of various vibrational modes. Utilizing laser cooling, it becomes possible to cool the ions so strongly that all vibrational modes are in their ground states. The ions also possess a lot of internal (energy) states. We take care that, like the vibrational modes, all the ions are also initially in their ground states. This situation will be the starting point for 'quantum engineering'.

The decisive question is how to affect the ions in the desired way. The tools the experimenter has at hand are laser pulses which can be tailored for special aims. This means, besides setting their frequency, their intensity and duration can be properly adjusted such as to produce, in particular, π or $\pi/2$ pulses (see Section 2.6.4). Moreover, the pulses can be focused so tightly that just one selected ion is addressed. With the help of the mentioned technique, two types of transition can be induced. (*i*) A laser pulse resonant with a transition from the ion's ground state $|g\rangle$ to an excited state $|e\rangle$ can be used to excite an individual ion to any superposition of $|g\rangle$ and $|e\rangle$. (*ii*) A properly detuned (blue-shifted) laser pulse can simultaneously excite an ion and one selected vibrational mode, thus also affecting the remaining ions. In this process just one laser photon is destroyed. However, its energy is too large to be swallowed by the ion alone. So there is a surplus energy that is taken by the vibrational mode, its occupation number (phonon number) thereby being enhanced from zero to unity. Naturally, a laser pulse of the described kind can also be used to induce the inverse process, i.e., it can de-excite both the ion and the vibrational mode.

When we allow, by properly setting the laser frequency, only for the excitation of just one excited level $|e\rangle$ which is long-lived (for instance, it is coupled to the ground state $|g\rangle$ through a quadrupole transition), we are effectively dealing with a two-level system. This is formally equivalent to the system representing the two degrees of freedom of photon polarization, the states $|g\rangle$ and $|e\rangle$ corresponding to the states of left- and right-handed circular polarization or linear polarization in two orthogonal directions. So we might feel challenged to entangle two ions we will label 1 and 2, after the example of polarization entanglement (cf. (2.36) and (4.7)). This goal has actually been achieved with the following three-step procedure (Roos *et al*. (2004)).

First step: focusing a properly detuned $\pi/2$ pulse with phase φ on ion 1, we excite the following state

$$\frac{1}{\sqrt{2}}(|g\rangle_1 |g\rangle_2 |0\rangle_{\text{vib}} - \text{ie}^{\text{i}\varphi} |e\rangle_1 |g\rangle_2 |1\rangle_{\text{vib}}),$$

where $|0\rangle_{\text{vib}}$ and $|1\rangle_{\text{vib}}$ are the ground state and the first excited state, respectively, of the chosen vibrational mode. Obviously, we have already managed to entangle the ion 1 with the vibrational mode. So what has to be done is to replace the vibrational mode with the second ion.

Second step: applying a resonant π pulse with phase $\pi/2$ to ion 2, we bring it into its excited state, thus producing the state

$$\frac{1}{\sqrt{2}}(|g\rangle_1 |e\rangle_2 |0\rangle_{\text{vib}} - \text{ie}^{\text{i}\varphi} |e\rangle_1 |e\rangle_2 |1\rangle_{\text{vib}}).$$

Third step: with the help of a detuned π pulse with phase $\pi/2$, we de-excite both the second ion and the vibrational mode. This means, the state $|e\rangle_2 |1\rangle_{\text{vib}}$ is transformed into $-|g\rangle_2 |0\rangle_{\text{vib}}$. The state $|e\rangle_2 |0\rangle_{\text{vib}}$, however, is not affected since there is no possibility of annihilating a vibrational quantum (phonon). Thus we arrive at the desired state

$$|\psi\rangle = \frac{1}{\sqrt{2}}(|g\rangle_1 |e\rangle_2 + \text{ie}^{\text{i}\varphi} |e\rangle_1 |g\rangle_2), \tag{4.9}$$

where we have omitted the common factor $|0\rangle_{\text{vib}}$, as we did with the ground states of all other vibrational modes from the very beginning.

Crucial to the experimental success by Roos *et al*. (2004) is the capability of switching the laser beam from one ion to another and to detune it properly in time intervals that are very short compared with the lifetime of the ion's excited state. This was achieved with the help of an electro-optical beam deflector and an acousto-optical modulator, respectively.

4.4 Quantum fluctuations

4.4.1 Origin of fluctuations

Fluctuation phenomena are well known from classical physics, especially thermodynamics. Usually they take place in systems that are formed of a huge number of microscopic constituents, such as atoms or molecules. For instance, a little mirror fastened to a torsion spring undergoes some kind of rotary

'Zitterbewegung' (jittering) as a result of the irregular bombardment with air molecules. This effect sets a limit to the measurement precision of a galvanometer.

Moreover, the discreteness of the electric charge – the existence of electrons – gives rise to voltage and current fluctuations. Actually, a great variety of such phenomena has been observed in radio engineering. They disturb reception and transmission of radio signals and have been termed noise. As examples, let me note the so-called Nyquist noise, that is, voltage fluctuations in resistors and conductors (in particular antennae) due to the thermal motion of the electrons, and the shot noise, i.e., fluctuations of the electric current in electronic tubes, arising from the statistical character of electron emission from the hot cathode.

Thinking of intrinsically quantum mechanical fluctuations, what will first come into one's mind are phenomena originating from the corpuscular aspect of light, i.e., the existence of photons. Investigating light with the help of photodetectors, one will find that the detector clicks (each of them indicating that just one photon has been absorbed) follow statistical distribution laws that depend on how the light was produced. In fact, there are many similarities between photon statistics and classical particle statistics.

Actually, quantum theory offers a much wider spectrum of fluctuation phenomena. We can say that fluctuations are unavoidably associated with *any* quantum system. This becomes obvious from the uncertainty relations that hold for any pair of conjugate variables. Hence it is impossible that both conjugate variables are simultaneously sharp. At least one of them, normally both, will exhibit fluctuations when an appropriate measurement is made. In particular, it follows from Heisenberg's uncertainty relation that the classical concept of an orbit (at any instant of time, both the position and the momentum of a particle take sharp values) cannot be upheld for electrons bound in atoms or molecules.

Let us consider, as an instructive example, the ground state of a harmonic oscillator (a particle subjected to a restoring force that is proportional to the displacement from a centre). While in the classical description the particle will take its lowest energy when it rests at the potential minimum; according to quantum theory the particle will still exhibit fluctuations of its position and its momentum, the so-called zero-point fluctuations, when it is in the lowest energy state. This effect appears still more striking when we associate it with the electromagnetic field. In fact, an electromagnetic field can be considered as an assembly of harmonic oscillators that correspond to the field modes (for instance, monochromatic plane waves with definite propagation directions and polarizations). Then the electric and magnetic field strengths fluctuate even when all the modes are in their ground states. This means that no photons are present, with certainty – the field is in the vacuum state. Clearly, this is at variance with classical theory in which the vacuum is completely void of electromagnetic fields.

In the following, we will study in more detail the implication of Heisenberg's uncertainty relation for a field mode.

4.4.2 Squeezing

As was already mentioned in Section 2.7.2, the quadrature components of a single-mode electromagnetic field satisfy the canonical commutation relation. As a consequence, they obey Heisenberg's uncertainty relation, and hence must both be fluctuating in realistic fields. This is really surprising from the classical point of view. Intuitively, one will expect that the fluctuations of the quadrature components are the same. This is so, in fact, for the coherent states (2.17) and also the vacuum state (which is formally a coherent state with $\alpha = 0$). Moreover, those states satisfy Heisenberg's uncertainty relation (2.7) (with $\hbar$ being suppressed) with the equality sign. So they are distinguished by minimal quantum noise. Nevertheless, one may ask the interesting question whether there exist in theory, or even might be produced, so-called squeezed states, that is, field states that show an imbalance between the fluctuations of two conjugate quadrature components, Δx and Δp. The answer is yes, and a possible production mechanism is degenerate parametric down-conversion.

Let me go into some details. We are dealing with the following nonlinear process: in a nonlinear crystal a strong pump wave generates a new wave at half the pump frequency. (Signal and idler coincide in the degenerate case we are considering.) From the quantum mechanical point of view, the elementary process is the decay of a pump photon into two signal photons. We will assume the pump wave to be so intense that its depletion due to signal amplification can be neglected. So, in this so-called parametric approximation, the amplitude of the pump wave remains constant and may be described by a given complex number. Let us first examine the process in classical terms. Then we characterize also the generated field by a complex amplitude A, and the equation of motion takes the form

$$\dot{A} = \kappa A^*, \tag{4.10}$$

where the effective coupling constant κ is proportional to the nonlinear susceptibility of the crystal and the pump amplitude. We assume κ to be positive, for convenience. Since, apart from a normalization factor, the quadrature components x and p are the real and imaginary part, respectively, of A (cf. the quantum mechanical analogues (2.42)), we may rewrite (4.10) as

$$\dot{x} = \kappa x, \quad \dot{p} = -\kappa p. \tag{4.11}$$

Obviously, these equations of motion predict that the quadrature component x grows exponentially in time, whereas the conjugate quadrature component p decreases exponentially. It is readily seen that the same holds true for the corresponding uncertainties, i.e., we have

$$(\Delta x)_t = e^{\kappa t}(\Delta x)_0, \quad (\Delta p)_t = e^{-\kappa t}(\Delta p)_0, \tag{4.12}$$

where the interaction time t has to be identified with the passage time of the pump wave through the crystal.

Since the equations of motion (4.11) are linear, we can immediately translate them into quantum mechanical language by interpreting x and p as observables (Heisenberg picture). Hence the results (4.12) are also valid in the quantum mechanical description. Obviously, they exhibit the squeezing effect very nicely. When we start from the vacuum state for the signal, the uncertainty Δp falls below the vacuum value, at the cost of an increase of Δx. (That it is just the quadrature component p that becomes squeezed is simply a consequence of assuming κ to be positive. Choosing κ as a complex number, you can, of course, squeeze any quadrature component x_Θ that we defined in Section 2.7.2.) The field produced in the described way was termed 'squeezed vacuum'. It is noteworthy that it follows from (4.12) that the product $\Delta x \Delta p$ is a constant of motion. This means, in particular, that Heisenberg's uncertainty relation is fulfilled with the equality sign for 'squeezed vacuum', since this is so for the initial state.

Lacking a classical counterpart, squeezing is an intrinsically quantum mechanical effect. It should be emphasized, however, that the stronger the squeezing the higher the intensity of the squeezed light must be. This is evident, since the strong fluctuations of the conjugate quadrature component imply strong fluctuations of the field energy, which also necessitates the average energy to be large. So strongly squeezed fields are by no means microscopic! In view of this fact, the term 'squeezed vacuum' is obviously misleading.

The first to produce squeezed light were Slusher *et al.* (1985). Actually, they used four-wave interaction as the generating mechanism, with sodium vapour as a nonlinear medium. Clearly, to demonstrate that what they had generated was really squeezed light (actually, 'squeezed vacuum'), they had to measure the fluctuations of various quadrature components. This was achieved with the help of the balanced homodyne detection technique. As explained in Section 3.2, the latter allows one to measure arbitrary quadrature components x_Θ by properly setting the phase of the local oscillator. Feeding the output into a spectrum analyzer, you measure the square root of x_Θ^2 averaged over a certain time interval. In the 'squeezed vacuum' case, the average over x_Θ vanishes so that the spectrum analyzer gives you directly the uncertainty Δx_Θ. Varying the angle Θ through changing the local oscillator phase, the researchers found, in fact, at a certain angle a minimum below the vacuum value, thus providing evidence of the squeezing effect. The vacuum value is, of course, independent of Θ, and it is readily measured by simply blocking the signal.

4.5 Amplifier noise

Quite generally, quantum mechanics states that a noiseless amplifier – a dream not only of experimentalists but also of theorists – cannot exist, for fundamental reasons. This statement is based on the following theoretical argument: focusing on light amplification, one might be tempted to describe the amplification process (in the Heisenberg picture) in close analogy with the classical description through the simple relation

$$\hat{a}^\dagger_{\text{final}} = g\hat{a}^\dagger_{\text{initial}}, \tag{4.13}$$

where $\hat{a}^\dagger$ is the photon creation operator and g (> 1) is the gain. This relation describes, in fact, noiseless amplification; however, it is not consistent with the basic commutation relation

$$[\hat{a}, \hat{a}^\dagger] = \mathbf{1}, \tag{4.14}$$

which is the equivalent of Heisenberg's commutation relation for the quadrature components (see (2.41) and (2.42)). Obviously, when (4.14) is fulfilled in the initial state, according to (4.13) this would not be so in the state after amplification. To get rid of this discrepancy (which is indeed drastic for large amplification) the theorist resorts to the following recipe which we actually employed in Section 3.5: add a noise operator on the right-hand side of (4.13). Its average value (expectation value) being zero, it has the task of ensuring the validity of the commutation relation (4.14) in the final state. So the occurrence of noise in any amplification process is, in fact, unavoidable. What remains to be done is to find out the physical origin of noise in concrete cases.

Let us consider first a maser or a laser amplifier. As the names suggest – maser (laser) stands for **m**icrowave (**l**ight) **a**mplification by **s**timulated **e**mission of **r**adiation – the primary goal of the researchers was low-noise amplification of radiation. The noise source is readily identified as spontaneous emission, which unavoidably accompanies stimulated emission from excited atoms or molecules. In the formalism, this follows from the fact that the transition matrix element $\langle n + 1 | \hat{a}^\dagger | n \rangle$ that is characteristic of the generation of an additional photon in the presence of n photons, has the value $\sqrt{n + 1}$ (see Section 8.1). Hence the transition probability is proportional to $n + 1$, which has to be interpreted such that n stands for 'true' stimulated emission and 1 for spontaneous emission of a photon. While in the first case the photon fits perfectly well into the coherent field already present, in the second case the photon is not affected by the field. In particular, its phase is random, and this is the origin of amplifier noise.

Another amplification scheme is provided by nondegenerate parametric down-conversion. Here, a coherent pump wave (laser field), together with a weak signal field to be amplified, is sent through a suitable nonlinear crystal. As was already pointed out in Section 2.6.5, the amplification process is necessarily accompanied by the generation of a second field, the so-called idler. It is just the idler that produces quantum noise. Let me show this in some detail. Actually, the Wigner formalism provides an elegant means to describe the interaction process in the parametric approximation. Then the equations of motion become, in fact, linear so that the evolution of the Wigner function becomes very simple, as was pointed out in Section 2.4.

We start from a state in which a low-intensity signal is already present, but no idler. So the initial Wigner function is the product of the Wigner functions for the signal and the idler, respectively. The latter is in the vacuum state whose Wigner function is known to be a Gaussian. Inverting the solution of the classical equations of motion, we can readily write down the Wigner function for the combined system: signal + idler (for details see Section 2.4). This total Wigner function will no longer be factorizable, thus indicating that signal and idler are entangled. We are, however, only interested in the signal, and hence trace over the idler (see Section 2.3). This tracing process proves to be a convolution with a Gaussian. So we arrive at a result quite similar to what we found in the analysis of simultaneous measurement of conjugate variables based on beamsplitting (see Section 3.5). Apart from a scaling transformation (obviously, the Wigner function becomes 'blown up' as a result of amplification), the Wigner function for the signal proves to be just the Q function for the initial signal (Leonhardt and Paul, 1993). This means that the original Wigner function for the signal gets smoothed, which, in fact, indicates the presence of noise.

It is interesting to note that an analysis of an ideal laser amplifier (approximately all atoms being excited) leads also to the Q function (Schleich, Bandilla and Paul, 1992). This shows that both a parametric amplifier and a laser amplifier produce the same amount of noise, which actually corresponds to the quantum mechanical limit.

5

Philosophy

5.1 Schrödinger's cat states

A prerequisite to getting quantitative information on the microscopic world is the existence of measuring apparatuses that indicate definite measuring results. The idea that a pointer of a measuring device is itself in an indeterminate state waiting for someone to make the wavefunction for the combined system (object + measuring apparatus) collapse by simply having a look on the apparatus, thus reading it out, appears absurd. I think, there is no doubt that measuring devices will behave like other objects in the macroscopic world in which there is no place for 'intrinsic' uncertainty.

Actually, this fact becomes questionable, at least it needs further theoretical investigation, when it is claimed that the quantum mechanical description applies also to macroscopic bodies, in particular measuring devices. (There are physicists who do not shrink back from attributing a single wavefunction even to the whole universe!) The dilemma originates from the quantum mechanical rule following from the linearity of the Schrödinger equation, that a pure quantum state remains a pure quantum state during its evolution. So, assuming that both the microscopic system to be observed and the measuring apparatus are initially in a pure state each, the combined system (object + measuring apparatus) will still be in a pure state after the interaction has taken place. Hence one will need 'someone' to bring about the collapse of the wavefunction. (In a realistic approach to this problem, the agent is recognized to be the environment with its virtually infinite number of degrees of freedom.) Schrödinger, being fully aware of this dilemma, illustrated it ingeniously through a fictitious 'diabolic device'. Let me cite from his seminal paper (Schrödinger, 1935):

One can even set up quite ridiculous cases. A cat is penned up in a steel chamber, along with the following diabolic device (which must be secured against direct interference by the cat): in a Geiger counter there is a tiny bit of radioactive substance, *so* small, that *perhaps* in the course of one hour one of the atoms decays, but also, with equal probability, perhaps none; if it happens, the counter tube discharges and through a relay releases a hammer which shatters a small flask of hydrocyanic acid. If one has left this entire system to itself for an hour, one would say that the cat still lives *if* meanwhile no atom has decayed. The first atomic decay would have poisoned it. The ψ-function of the entire system would express this by having in it the living and the dead cat (pardon the expression) mixed or smeared out in equal parts.

The Schrödinger's cat paradox became famous among physicists. However, while he clearly spoke of the wavefunction for the *entire* system, it became a widespread belief among theorists that Schrödinger had supposed the cat to be in a superposition state

$$|\psi\rangle_{\text{cat}} = \alpha \, |\text{alive}\rangle + \beta \, |\text{dead}\rangle , \qquad (5.1)$$

where α and β are complex coefficients satisfying the normalization condition $|\alpha|^2 + |\beta|^2 = 1$. This is, of course, quite another thing!

Nevertheless, the misunderstanding, as it probably was, proved to be very fruitful. It stimulated two questions that can actually be tackled experimentally. (*i*) Since superposition states are at the heart of quantum theory, one may ask whether they might exist, i.e., can be produced experimentally, in the diffuse transition region between microcosm and macrocosm, that is, the mesoscopic domain. (*ii*) How stable are such states (that were readily baptized 'Schrödinger's cat states')?

Actually, both questions were answered in a pioneering experiment by Brune *et al.* (1996). Starting from a low-intensity microwave cavity field in a Glauber state $|\alpha\rangle$ with an average photon number varying from 0 to 10, they managed to bring this state into the superposition state

$$|\psi\rangle_{\text{cat}} = \frac{1}{\sqrt{2}} \left(\left|e^{i\varphi}\alpha\right\rangle + \left|e^{-i\varphi}\alpha\right\rangle \right), \qquad (5.2)$$

where φ is a constant phase. To this end, they sent a rubidium atom prepared, with the help of an intense coherent microwave field, in a superposition of two circular Rydberg states, through the cavity containing the microwave field to be altered. The dispersive (nonresonant) interaction between the atom and the field resulted in an entangled state for the combined system. To get rid of the entanglement, the atom, after having left the cavity, was made to traverse a second intense microwave field and afterwards was detected with the help of a field ionization detector. When it is found in the lower level, the field is known from theory to be in the desired state (5.2). (For details, see Schleich (2001).)

In a subsequent experiment the authors investigated the so-called decoherence of this state, that is, the decay of the correlation between the two components in (5.2). Surprisingly, they found that this decay is much faster than the damping of the field. Moreover, they observed that the decay time was decreasing with the distance, in phase space, between the two components. So a 'cat state' is the more fragile, the better its components can be distinguished. This gives us a hint as to how to understand that such states actually do not exist in the macroscopic domain. If it were possible to produce them, they would immediately decay due to their inevitable interaction with the environment.

5.2 The EPR paradox

In 1935 Einstein, jointly with Podolsky and Rosen (the names being soon abbreviated to EPR), asked the intriguing question 'Can quantum mechanics be considered complete?' (Einstein, Podolsky and Rosen, 1935). What they put forward was indeed a severe criticism aiming at the heart of quantum theory. They based their argument on a gedanken experiment to be carried out on an entangled system. They considered two particles with coordinates x_1, x_2 (one-dimensional, for simplicity) and momenta p_1, p_2, that are coupled such that *both* the sum of their momenta, S, and the difference of their positions, D, have sharp values. Since the variables S and D commute, such states do exist in quantum theory. They described the whole system by the delta function $\delta(x_1 - x_2 - a)$, which is, in fact, an eigenfunction of both $D = x_1 - x_2$, with eigenvalue a, and $S = p_1 + p_2$, with eigenvalue 0. First, they supposed a position measurement to be made on the first particle. Owing to the entanglement, this gives us simultaneously the position of the second particle. Assuming the two particles to be far away from one another, one can say that the measurement could not have affected, in a physical way, the second particle, since any action propagates at a velocity that does not exceed that of light. A physical action that might originate from the measurement would definitely come too late. So one is led to the conclusion that the position of the second particle, x_2, had the indirectly measured value already before the measurement on the first particle. This would mean that x_2 is a *real*, in the sense of classical 'objective reality', property of the second particle, or, as EPR put it more cautiously, that, 'There exists an element of physical reality corresponding to this physical quantity.'

Things become strange, however, when we continue with the following reasoning. Instead of the position, you *could* have measured the momentum of the first particle, and this *would* have also given you the momentum of the second particle. Arguing along the same lines as before, one would conclude that the momentum of the second particle, p_2, was already sharp before the measurement.

However, the 'objective reality' of both x_2 and p_2 is in conflict with Heisenberg's uncertainty relation. Since the latter is a consequence of the commutation relations that are basic to quantum mechanics, the contradiction is indeed severe. So it seems that EPR are right in claiming that the quantum mechanical description is incomplete since it asserts that certain variables are indeterminate which are sharp 'in reality'. This argument became famous as the EPR paradox, giving rise to never-ending, as it seems, discussions on the interpretation of quantum mechanics.

Actually, there is no paradox felt by an experimentalist. On an individual two-particle system you can measure either x_1 or p_1, which allows you to predict either x_2 or p_2. A contradiction to quantum theory is found only when the predictions of a real measurement are compared with those of a fictitious measurement that has never been – actually, could not have been – performed. From the viewpoint of quantum theory, this is not physics, and hence the EPR paradox belongs to the realm of philosophy.

A quantum theorist will not see a paradox either. He will agree with EPR that a measurement of x_1, for instance, makes it possible to predict, with certainty, the value of x_2. However, he will not follow them in concluding – from the argument that the measurement cannot have exerted a physical action on the second particle – that this value existed, as a real property of the second particle, already before the measurement. Instead, he will say that x_1 and x_2 are both indeterminate before the measurement. However, the two particles are interconnected in a subtle way such that they are bound to 'respect' the constraint $x_1 - x_2 = a$, which actually becomes manifest in the measurement.

It will be helpful to consider pairs of polarization-entangled photons. In Section 4.3.1 we learned how the correlations are produced. This happens in the emission process (the two-photon cascade), and the correlations have their origin in the conservation law for angular momentum. So it appears natural that they persist while the photons propagate. From (4.8) it becomes obvious that a polarization measurement on the first photon brings, via reduction of the wavefunction, the second photon in a polarization eigenstate too, from which the outcome of a corresponding measurement can indeed be predicted with certainty. However, this was not so before the first measurement!

It should be emphasized that the 'interconnectedness' of two (or more) systems we are speaking about is an intriguing quantum feature that has no counterpart in classical physics (which makes a 'physical understanding' difficult). Formally, it requires the whole system to be described by a common (nonfactorizable) wavefunction. This forbids us to conceive of the subsystems as being independent. The whole system has properties that cannot be revealed from an analysis of its parts. This is meant by the phrase 'the whole is more than the sum of its parts', and we learn that quantum mechanics is a manifestly nonlocal theory.

In 1951, a first step towards a practical realization of an EPR experiment was taken by D. J. Bohm who considered a pair of spin-$\frac{1}{2}$ particles being in a state with total spin 0 (s state). Later on, this proved to be a lucky idea, since it inspired

J. S. Bell, a theorist at CERN, to analyze EPR correlations in terms of hidden variables. His striking results, in turn, stimulated experimental work on polarization-entangled photon pairs. In fact, the polarization properties of photons resemble closely the spin properties of spin-$\frac{1}{2}$ particles.

So let us focus on the correlated photon pairs we have studied already in Sections 2.6.5 and 4.3.1. In what follows, we will specialize to an entangled state of the form (4.8). Then we can get information on one photon of a pair, say that labelled 2, by observing the other photon, labelled 1, that may be far away. We will name the observers dealing with photon 1 and photon 2 Alice and Bob, respectively. Let Alice measure linear polarization. This can be done with the help of a polarizing prism, with a detector in each of its output channels. Let us assume that the prism is oriented such that it splits an incoming beam into two beams that are polarized in the x and y direction, respectively. Then Alice will find 'her' photon to be either x or y polarized. Say, it is x polarized in a given case. It follows from the entanglement (formally from the 'reduction of the wavepacket' (4.8)) that Bob's photon has immediately 'become' x polarized too.

Alternatively, Alice may rotate her prism, thus being enabled to measure linear polarization along the axes of a rotated frame, say x' and y'. In this case Bob's photon will immediately 'become' x' or y' polarized, dependent on the outcome of Alice's measurement. Moreover, she may also decide to measure circular polarization. To this end, she needs only insert a quarter-wave plate in front of the polarizing prism. This will 'make' Bob's photon circularly polarized too, according to (4.7). All this looks like 'spooky action at a distance', as Einstein had put it.

One might be tempted, and some physicists actually were, to ask whether this strange phenomenon could be utilized to transmit signals with velocity larger than that of light. However, this would manifestly violate causality which is, in fact, one of the pillars of the edifice of modern physics. So such a project is bound to fail, from the very beginning. Nevertheless, from a thorough criticism of those attempts to outwit the approved principle of causality, one learns a lot about quantum physics. This will become evident in the next section.

5.3 Causality and nonobjectifiability

A characteristic feature of an EPR experiment is that an observer must leave to chance what will be the outcome of the measurement. So Alice cannot encode a message she wants to send to Bob using the convention that, for instance, x polarization stands for '1' and y polarization for '0'. However, there is another possibility. What Alice can do is to decide, at will, whether she will measure circular or linear polarization. She needs only insert a quarter-wave plate in front of the polarizing prism, or remove it. So the code will be: circular polarization (whereby it remains open whether it is right- or left-handed) means, say '1', and linear polarization (without specifying whether it is x- or y-polarization) '0'. When Alice and Bob are far away from one another, a message encoded in this way could, in fact, be transmitted with superluminal velocity, provided Bob is able to decode it! His task would be to distinguish a linearly polarized photon from a circularly polarized one. This, however, is impossible! We assume that Bob has been informed by Alice beforehand what her set-up looks like and how she plans to operate it. Using the same apparatus (with the polarizing prism being oriented as Alice's) he has the freedom, like Alice, to insert a quarter-wave plate, or to remove it. In the first case he will find his photon to be either right-handed or left-handed circularly polarized, and in the second case he measures either x- or y-polarization. So, which kind of polarization he observes is *solely* determined by the choice of his set-up, and hence does not give him any information on the 'true' polarization of his photon. Paradoxically, Alice knows it precisely from her measurement, but she cannot communicate it to him so fast.

The question is, can Bob do better? An idea might be to clone the photon. This would indeed resolve his problem, since by repeated cloning he would get an ensemble of photons whose polarization state is readily detected: he will split this ensemble (arbitrarily) into two, and measure linear polarization on one and circular polarization on the other. In one of the two cases he will find all photons to be equally polarized, and this gives him the 'true' polarization of the initial photon. This sounds good; however, cloning was shown to be impossible, since it is in conflict with the linearity of quantum theory. In fact, a 'no-cloning theorem' was proved by Wootters and Zurek (1982).

But maybe amplification can help us? A polarization-insensitive quantum amplifier would produce an intense field, which is virtually a classical field, from a single photon. The polarization of the amplified field is readily measured with techniques well known from classical polarization optics. There is, however, a great drawback to amplification, namely the inevitable amplifier noise (cf. Section 4.5). One might ask, however, whether *all* information on the polarization of the original photon gets lost in the amplification process. Wouldn't we intuitively expect that the original photon's polarization leaves a 'trace' in the amplified field, that might be recovered? Seemingly, this guess is supported by a quantum mechanical treatment of the amplification process. Certainly, the wavefunction for the amplified field will depend on the polarization of the initial photon. However, it must be remembered that the quantum mechanical description always applies to an *ensemble* of systems! It does not tell you how an individual system will behave.

So our attempts to beat the causality principle have failed. Information transmission with superluminal velocity cannot be achieved. This is due to the paradoxical situation that Alice can be sure that Bob's photon is in a well-defined polarization state known to her, but Bob is unable, in principle, to find this out. Actually, such a strange situation is not restricted to entangled systems. We encounter it, for instance, when a researcher prepares a system in an eigenstate of a certain observable. Then he knows that the latter has a well-defined value. However, an independent observer can by no means recover this information. This again shows up the nonobjectifiability of quantum mechanics: contrary to classical physics, one cannot consider quantum properties as being 'real', in the sense of 'objective reality'.

But we can also turn the tables. Being convinced that quantum mechanics is in full agreement with the causality principle, we may conclude from the discussion of EPR experiments that cloning *must be* impossible and that a (polarization-insensitive) amplifier *must* work in such a way that no information on the initial photon's polarization can be extracted from the amplified field in any individual case. To my mind, those connections between quite different issues are really amazing. In particular, I am excited by the fact that quantum mechanical nonobjectifiability is not a mere whim propagated by theorists, but actually saves causality.

5.4 Protection against eavesdropping

When secret messages are to be communicated, the safety of the code is of vital importance. Absolute security is guaranteed when a set of random numbers is used as a key for encoding. This is so in the so-called Vernam code, where the text written in binary digits '0' and '1', and the key given by a random sequence of binary digits (which must have the same length as the text) are added modulo 2, digit by digit. Decoding is, of course, possible only with a copy of the key. So the task is to provide both the sender (Alice) and the receiver (Bob) with an identical sequence of random numbers, without giving an eavesdropper a chance to learn it without being detected.

Quantum theory offers a scheme to achieve this goal. As in the teleportation experiment explained in Section 3.6.4, the members of polarization-entangled photon pairs are sent to Alice and Bob. Alice and Bob each have the same measuring device at their disposal, namely a polarizing prism with a detector in each output channel, that can be oriented either parallel to the vertical or at an angle of 45°. For any orientation of the prism, an observer will detect 'his' (or 'her') photon in one of the two output channels, which gives him (or her) a random sequence of binary digits. The point is that such a sequence is the same for Alice and Bob, owing to the entanglement (see Sections 2.6.5 and 4.3.1), *provided* they have set their apparatuses in the same way, in any individual measurement. What shows up here is just EPR correlations.

To protect their data from eavesdropping, they will adopt the following strategy: independently from one another, they change the orientation of their prisms randomly. After completion of their measurements, Alice informs Bob, via a *public transmission channel*, which settings she chose for the first, second, ... measurement. This gives Bob an opportunity to find out, by comparison, in which cases the settings coincided. He communicates his result to Alice, and they both select those data that correspond to such coincidences, eliminating the rest. This procedure leaves them with *identical* data sets, as required.

Let us now assume that an eavesdropper (this term almost inevitably makes one think of a lady named Eve) is at work, observing the photons that are on their way to, say, Bob. Measuring their polarization meets, however, with two difficulties: (*i*) Eve does not know the correct settings for the individual measurements. When she chooses the false orientation, she will observe the correct (namely Alice's) result in only 50 per cent of the cases. (*ii*) In her measurement, the photon inevitably gets lost. So she has to replace it with a new one that has the polarization she has measured. This will, however, destroy the EPR correlation with Alice's photon, in the case where she has chosen the false orientation. This will lead to discrepancies that will show up when Alice and Bob carry out a routine check: they will select, after a key they had agreed on beforehand, certain data from their sequences and compare them. (Since this happens in public, those data have to be eliminated afterwards, anyway.) Thus Eve cannot avoid detection, in principle. So EPR correlations might actually be of practical relevance. It should be noticed, however, that data transmission via photons has a serious handicap: the transmission rate is extremely low.

5.5 EPR correlations and hidden variables

Let us come back to Bohm's version of an EPR experiment. He imagined two spin-$\frac{1}{2}$ particles with sharp value 0 of the total spin, to propagate in given opposite directions. (One might think of a diatomic molecule being in a singlet state, that has been cautiously made to dissociate.) As in the case of polarization-entangled photon pairs studied before, we assume two observers, Alice and Bob, to perform independent measurements on the particles 1 and 2, respectively. Being endowed with Stern–Gerlach apparatuses assumed to be equally oriented, they can measure the spin components along a fixed direction (orthogonal to the propagation directions and indicated by a unit vector *a*), $\sigma_a^{(1)}$ and $\sigma_a^{(2)}$. The possible measured values are +1 or −1 (in units of $\hbar/2$). Performing repeated experiments, Alice and Bob will find a random number series each. This will be so, say for Bob, irrespective of whether Alice is measuring, or what she might have chosen to measure (e.g., she could have rotated her measuring apparatus), or whether she is present at all. However, when they both have measured with their Stern–Gerlach apparatuses

equally oriented and meet afterwards to compare their data, they might be surprised
to see that those are strictly correlated in the form

$$\sigma_a^{(1)} = -\sigma_a^{(2)}, \tag{5.3}$$

which is typical of an EPR experiment.

Bohm's version differs favourably, however, from the original EPR experiment.
First, we are dealing now with discrete, rather than continuous, variables. The most
important point, however, is that Bohm's system has greater experimental potential.
Indeed, Bob can orient his Stern–Gerlach magnet in a different way from Alice.
So he can measure the spin along a different direction, say *b*. This opens a way
to measure specific correlation *functions*. Owing to the rotational symmetry of
the system, they will depend on the relative orientation of the two Stern–Gerlach
magnets only, i.e., the angle between *a* and *b*.

Bell (1964) had the ingenious idea of making those correlation functions the
basis for a comparison of quantum mechanics with hidden-variables theories that
answered to the EPR criticism of quantum theory. Indeed, the introduction of hidden
variables seems to be necessary to overcome the alleged incompleteness of quantum
theory. Moreover, a hidden-variables theory can be conceived such that the causality
principle is obeyed in the sense that what happens to one subsystem is in no way
affected by what may be done with the other. This is just what Einstein considered
as an indispensable postulate. It makes the theory local.

The question Bell actually asked was: 'Are the quantum mechanical predictions
for the correlation functions compatible with the implications of deterministic local
hidden-variables theories?' This is indeed an unusual question, since in former
studies it was always required that hidden-variables theories, of course, should
reproduce all quantum mechanical predictions. However, Bell came to the amazing
conclusion that the whole class of deterministic hidden-variables theories is in
conflict with quantum theory.

So let me explain what Bell showed. He started from the concept that hidden
variables λ predetermine the outcome of any measurement. The symbol λ may stand
for discrete or continuous variables, or more abstract mathematical objects. What
matters only is that (positive-definite) probability densities $\rho(\lambda)$ exist that fulfil
the familiar normalization condition $\int \rho(\lambda) d\lambda = 1$. Here, the integral indicates
a summation over the contributions from all values of λ. Such a probability
distribution is the counterpart of the quantum mechanical wavefunction. It describes
an ensemble of equally prepared systems, whereby the role of the quantum
mechanical expectation value of an observable is now played by an average over
a function $f(\lambda)$, $\int f(\lambda) \rho(\lambda) d\lambda$, whose values are the predetermined values of the
observable. Here, it is supposed that any *individual* system is fully characterized

by a certain value λ. Though we cannot know it – it is hidden by definition – the mere assumption that it exists has interesting consequences, as we will soon see.

What shall be analyzed is correlations that show up in simultaneous measurements of the spin components along two directions a and b, $\sigma_a^{(1)}$ and $\sigma_b^{(2)}$. Bell chose as a correlation function the ensemble average over the product of those spin components, $E(a, b)$. Owing to the rotational symmetry of the system, this function will depend on the angle between a and b only. In the quantum mechanical description this correlation function is the corresponding expectation value, which is readily calculated from the well-known wavefunction for an s state to be

$$E_{\mathrm{qu}}(a, b) = \left\langle \sigma_a^{(1)} \sigma_b^{(2)} \right\rangle = -a \cdot b. \tag{5.4}$$

Experimentally, the correlation function in question will be obtained by measuring the spin components separately, multiplying them and forming the average over a great number of experiments. Since a measurement of a spin component can yield only $+1$ or -1, only four different outcomes of a single experiment are possible: (i) $\sigma_a^{(1)} = +1, \sigma_b^{(2)} = +1$; (ii) $\sigma_a^{(1)} = +1, \sigma_b^{(2)} = -1$; (iii) $\sigma_a^{(1)} = -1, \sigma_b^{(2)} = +1$; (iiii) $\sigma_a^{(1)} = -1, \sigma_b^{(2)} = -1$. Accordingly, the measured correlation function can be decomposed into four terms

$$E(a, b) = w_{++} - w_{+-} - w_{-+} + w_{--}, \tag{5.5}$$

where w_{++} denotes the probability for the event (i), and so on. Since a spin measurement requires the detection of the particle in one of the two output channels of the Stern–Gerlach magnet (cf. Section 3.1.1), the ws are, in fact, normalized coincidence counting rates.

Let us now describe the correlations in terms of a deterministic local hidden-variables theory. Determinism means that the outcomes of individual spin measurements are uniquely predetermined by the actual value of λ, and locality requires that a measurement on one particle is in no way affected by a simultaneous measurement on the other. So we can postulate the existence of functions $A(a, \lambda)$, $B(b, \lambda)$ with the following properties. They take values $+1$ and -1 only, which are the possible results of a measurement of $\sigma_a^{(1)}$ or, likewise, $\sigma_b^{(2)}$. The crucial point is that each of them depends, apart from λ, only on the setting of the apparatus that actually acts on the particle. Then the correlation function can be written as

$$E(a, b) = \int A(a, \lambda) B(b, \lambda) \rho(\lambda) d\lambda. \tag{5.6}$$

As was mentioned above, quantum mechanics predicts that for parallel directions, $a = b$, the outcomes of the two spin measurements are opposite in sign. Since this is a basic physical feature – it is an immediate consequence of the assumption that the total spin of the system is zero – it should also be preserved in the hidden-variables theory. Hence we postulate the relation

$$A(\boldsymbol{a}, \lambda) = -B(\boldsymbol{a}, \lambda) \tag{5.7}$$

to hold true for all directions $\boldsymbol{a}$ and all values of λ.

Of course, one cannot derive, from the general postulates, a quantitative expression for the correlation function. This makes a direct comparison with the quantum mechanical prediction impossible. But maybe the correlations for different settings of the Stern–Gerlach magnets are subjected to certain restraints that might be used for checking? In fact, Bell (1964) was lucky to find such a condition. He considered the difference of two correlation functions, and with the above premises he needed only a few lines of simple mathematical reasoning to derive the constraint

$$|E(\boldsymbol{a}, \boldsymbol{b}) - E(\boldsymbol{a}, \boldsymbol{c})| \leq 1 + E(\boldsymbol{b}, \boldsymbol{c}), \tag{5.8}$$

which became famous as the first Bell inequality.

It is really of great importance since it allows for a comparison with the quantum mechanical prediction (5.4). It may be felt as a surprise that the latter is actually in conflict with Bell's inequality. This becomes evident when the (coplanar) vectors $\boldsymbol{a}$, $\boldsymbol{b}$ and $\boldsymbol{c}$ are chosen such that $\boldsymbol{c}$ makes an angle of $2\pi/3$ with $\boldsymbol{a}$, and $\boldsymbol{b}$ makes an angle of $\pi/3$ with both $\boldsymbol{a}$ and $\boldsymbol{c}$. Then we have $\boldsymbol{a} \cdot \boldsymbol{b} = \boldsymbol{b} \cdot \boldsymbol{c} = \frac{1}{2}$ and $\boldsymbol{a} \cdot \boldsymbol{c} = -\frac{1}{2}$. Thus the left-hand side of (5.8) takes the value 1, while the right-hand side becomes $\frac{1}{2}$, which constitutes a manifest contradiction.

So, which theory is right, quantum theory or hidden-variables theory? The exciting result of Bell's investigations is that this question might be answered experimentally. In fact, this prospect inspired researchers to devise and analyze theoretically realistic EPR experiments. Along with those studies, new forms of inequalities were found. A remarkable progress was that a generalized inequality, which became known as the second Bell inequality, was derived without requiring the theory to be deterministic. Instead, it can also have a stochastical character. This means that what happens with the detectors follows (unknown) statistical rules (depending on λ and the setting of the respective apparatus). What is important only is that locality is still ensured in the form that probabilities for given detectors to respond are independent when the detectors belong to different measuring apparatuses.

A suitable system for an experimental test was found in polarization-entangled photon pairs generated in an atomic cascade transition (see Section 4.3.1). Actually, there is a perfect analogy to Bohm's gedanken experiment, when polarizing prisms, combined with photodetectors, are used for measurement. In the experiment, use is made of the freedom to rotate the prisms independently

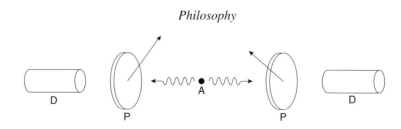

Fig. 5.1 Experimental set-up for the measurement of polarization correlations on two photons emitted in a cascade transition. A = atom; D = detector; P = polarizer; the arrows indicate the directions of polarizer transmittance.

around the propagation directions of the photons. A polarizing prism splits incident light into two partial beams that are mutually orthogonally polarized. Let the polarization directions be x_1, y_1 for the first prism, and x_2, y_2 for the second. Detectors are placed in the four output channels. So four coincidence counting rates $R(x_1, x_2), R(x_1, y_2), R(y_1, x_2), R(y_1, y_2)$ can be measured which, after normalization, are the precise equivalent of the ws in (5.5). So a correlation function can be defined after the example of (5.5), and determined experimentally. (It should be noticed that the directly measured coincidence counting rates contain contributions from random coincidences, which must be determined separately and subtracted.)

Actually, the first experiments followed a simplified scheme with polarizers being used instead of polarizing prisms (see Fig. 5.1). Nevertheless, the researchers were able to demonstrate violations of the second Bell inequality (for details see, e.g., Clauser and Shimony (1978)). The most impressive proof of such violations was given by Aspect, Grangier and Roger (1982), who realized the optimal scheme with two polarizing prisms, described above. They provided convincing evidence against local hidden-variables theories, thereby confirming the quantum mechanical predictions with high accuracy. So quantum theory comes off as a winner and we have to live with quantum mechanical peculiarities, such as nonlocality.

6

Interaction

The main task of a quantum theorist is to describe interaction processes, since they underlie most of the quantum phenomena we observe. Basically, there are two types of interaction: (*i*) interactions mediated by time-independent forces between (mostly elementary) particles that give rise to stable configurations such as atoms, molecules or nuclei and (*ii*) disturbance of quantum systems through short-time coupling to other systems, or external fields, which results in observable 'events'. In the following, I would like to give you some impression of how quantum mechanics deals with those problems.

6.1 Atomic structures

6.1.1 Schrödinger equation

In atomic and molecular systems it is the Coulomb attraction exerted from positively charged nuclei on electrons that makes possible stable structures with well-defined discrete energies. According to the general quantum mechanical formalism explained in Chapter 2, one finds the energy levels E_n, together with the corresponding states $|\psi_n\rangle$, as solutions of the eigenvalue problem for the energy operator (Hamiltonian) $\hat{H}$

$$\hat{H}\,|\psi_n\rangle = E_n\,|\psi_n\rangle. \tag{6.1}$$

The energy of a particle system is the sum of the kinetic energies of the particles and the potential energy. The kinetic energy of a particle of mass m is $p^2/2m$, where p is the momentum, and the potential is a function of the coordinates. So the eigenvalue problem (6.1) is readily formulated in the Schrödinger representation. Then, an eigenvector $|\psi_n\rangle$ becomes a complex function ψ_n of the particles' coordinates, and the momentum operators are represented by the differential operators, with respect to the coordinates $x_1^{(j)}, x_2^{(j)}, x_3^{(j)}$ of the jth particle, $\hat{p}^{(j)} = (\hbar/\mathrm{i})\mathbf{grad}^{(j)}$. It should be noted that, when k (> 1) particles are involved, the wavefunction is defined in an abstract $3k$-dimensional configuration space which only in the case $k = 1$ becomes identical to the space we live in.

After what has been said, the task of solving (6.1) in the Schrödinger representation, that is, the time-independent Schrödinger equation, amounts to solving a second-order linear partial differential equation. So we can benefit from the well-established mathematical theory of partial differential equations. And how does energy quantization come about? The answer is that the wavefunction is required to be normalizable. In other words, an eigenfunction must satisfy the boundary condition that, when going to infinity, it falls to zero strongly enough so that the integral of its squared absolute value taken over the configuration space is finite. Moreover, the eigenfunctions are required to be uniquely defined.

6.1.2 Hydrogen atom

For illustration, let us briefly discuss the hydrogen atom. This simplest atomic system, one can imagine, played a crucial role in the development of quantum mechanics as an ideal test object. In fact, on the one hand, the energy eigenvalue problem could be rigorously solved (1925 by Pauli on the basis of Heisenberg's commutation relations and 1926 by Schrödinger in his 'wave-mechanical' approach to quantum theory), and, on the other hand, precise spectroscopic data were available for comparison. The excellent agreement between theory and experiment showed that the quantum mechanical concepts were sound.

A hydrogen atom consists of an electron that is moving in the Coulomb field of a proton. Since a proton is much heavier than an electron (the proton mass is about 1836 times the electron mass), we can neglect the back action of the electron on the proton. Moreover, we idealize the proton as a point-like particle resting at the origin of the coordinate system. Hence the electron moves in a given central field, the Coulomb field, which gives rise to the potential

$$V(r) = -\frac{e^2}{r}, \tag{6.2}$$

where e is the elementary electric charge measured in electrostatic units, and r is the distance from the centre. Since the potential does not depend on the electron spin, we can disregard the latter in the following analysis. The operator for the electron's kinetic energy, on the other hand, reads

$$\hat{T} = -\frac{\hbar^2}{2m}\Delta, \quad \Delta = \frac{\partial^2}{\partial x_1^2} + \frac{\partial^2}{\partial x_2^2} + \frac{\partial^2}{\partial x_3^2}, \tag{6.3}$$

where m is the electron mass and Δ is known as the Laplace operator. So the Schrödinger equation can be written as

$$\left[\Delta + \frac{2m}{\hbar^2}\left(\frac{e^2}{r} + E\right)\right]\psi = 0, \tag{6.4}$$

where we have suppressed the subscript n.

Since the problem is spherically symmetrical, it will be advantageous to introduce polar coordinates r, ϑ, φ. Then the Laplace operator takes the form

$$\Delta = \frac{1}{r^2}\frac{\partial}{\partial r}\left(r^2\frac{\partial}{\partial r}\right) + \frac{1}{r^2}\Lambda, \tag{6.5}$$

where the operator Λ stands for

$$\Lambda = \frac{1}{\sin\vartheta}\frac{\partial}{\partial\vartheta}\left(\sin\vartheta\frac{\partial}{\partial\vartheta}\right) + \frac{1}{\sin^2\vartheta}\frac{\partial^2}{\partial\varphi^2}. \tag{6.6}$$

The shape (6.5) of the Laplace operator makes it possible to separate the angular dependence of the wavefunction in the form

$$\psi(x_1, x_2, x_3) = u(r)f(\vartheta, \varphi), \tag{6.7}$$

were $f(\vartheta, \varphi)$ is bound to satisfy the equation

$$\Lambda f(\vartheta, \varphi) = \text{const} \cdot f(\vartheta, \varphi). \tag{6.8}$$

Actually, the operator Λ has a physical meaning: multiplied by $-\hbar^2$, it is the square of the angular momentum operator $\hat{\boldsymbol{L}} = \hat{\boldsymbol{x}} \times \hat{\boldsymbol{p}}$ written in polar coordinates. The eigenfunctions of $\hat{\boldsymbol{L}}^2$ are known to be the spherical harmonics

$$Y_l^m(\vartheta, \varphi) = N_{lm}P_l^{|m|}(\cos\vartheta)e^{im\varphi}, \tag{6.9}$$

where the functions $P_l^{|m|}$ are associate Legendre polynomials and N_{lm} are normalization constants. The corresponding eigenvalue is $\hbar^2 l(l+1)$, where l can take the values $l = 0, 1, 2, \ldots$ Moreover, the spherical harmonics are eigenfunctions of the x_3 component (the projection on the polar axis) of the vector $\hat{\boldsymbol{L}}$, $\hat{L}_3 = -i\hbar\partial/\partial\varphi$, with eigenvalues $m\hbar$ $(m = -l, -l+1, \ldots, l)$.

So we can identify $f(\vartheta, \varphi)$ with a spherical harmonic, and (6.8) becomes

$$\Lambda Y_l^m = -l(l+1)Y_l^m. \tag{6.10}$$

Thus the Schrödinger equation (6.4) reduces to the ordinary differential equation

$$\frac{d^2u}{dr^2} + \frac{2}{r}\frac{du}{dr} + \frac{2m}{\hbar^2}\left[E + \frac{e^2}{r} - \frac{\hbar^2 l(l+1)}{2mr^2}\right]u = 0. \tag{6.11}$$

Here, the term proportional to $l(l + 1)$ can be interpreted as the potential corresponding to the centrifugal force. This can be seen from comparison with the classical motion of a particle along a circle.

So we have already found an important physical result. A hydrogen atom, being in an energy eigenstate, has a definite (quantized) angular momentum characterized by the quantum numbers l and m. It should be stressed that we arrived at this result without making use of the special form of the potential (6.2). Hence it applies to the general case of a spherically symmetrical potential.

The normalizable solutions of (6.11) are found to be a (decreasing) exponential multiplied by a polynomial. (Actually, the polynomials are obtained from an infinite power series by requiring the latter to break off in order to make it convergent.) The corresponding energy eigenvalues have the remarkably simple form

$$E = -\frac{me^4}{2\hbar^2} \frac{1}{(n_r + l + 1)^2}, \tag{6.12}$$

where $n_r = 0, 1, 2, \ldots$ is the so-called radial quantum number. Since the energy levels depend only on the sum $n_r + l + 1$, the latter was declared the principal quantum number n, $n = n_r + l + 1$, which can take the values $n = 1, 2, 3, \ldots$ Thus we can write

$$E_n = -\frac{me^4}{2\hbar^2} \frac{1}{n^2} \tag{6.13}$$

instead of (6.12). The corresponding eigenstates are characterized by the quantum numbers n, l and m. This means that for $n > 1$ the energy levels are degenerate. For a given principal quantum number n, the angular momentum quantum numbers $l = 0, 1, \ldots, n-1$ are allowed. An additional degeneracy is caused by the fact that a (quantized) angular momentum vector corresponding to a quantum number l can be oriented in $2l + 1$ different ways, as indicated by the quantum number m, the so-called magnetic quantum number, which can take the values $m = -l, -l+1, \ldots, l$. So the degeneracy of an energy level with principal quantum number n is n^2. Taking into account the electron spin, which has two possible orientations described by the quantum number s ($= -1/2$ or $1/2$), with respect to the quantization axis, one sees that the degeneracy is actually $2n^2$.

The result (6.13) tells us that the energy values are negative. The deepest level has the principal quantum number $n = 1$, and for growing n the distance between neighbouring levels becomes smaller and smaller. For $n \to \infty$ the energies tend to the value 0, which marks the ionization threshold. For positive energies, the electron is no longer bound but is able to move away from the nucleus.

With the help of (6.13) and Bohr's frequency condition (1.3) we readily recover the observed series of emission lines

$$\nu_{kn} = \frac{me^4}{4\pi\hbar^3}\left(\frac{1}{n^2} - \frac{1}{k^2}\right) \tag{6.14}$$

(n fixed, $k = n+1, n+2, \ldots$). Those spectral series were named after researchers: Lyman ($n = 1$; ultraviolet spectrum), Balmer ($n = 2$; visible spectrum), Paschen ($n = 3$), Brackett ($n = 4$), and Pfund ($n = 5$), the last three series lying in the infrared region.

We have seen that the energy states of a hydrogen atom are characterized by the quantum numbers n, l and m. The following notation became customary. To specify the angular momentum, small letters are used: $s, p, d, f, g, \ldots$ stand for $l = 0, 1, 2, 3, 4, \ldots$, respectively. Moreover, the actual value of the principal quantum number is written in the first place. So, for instance, a state with $n = 4$ and $l = 1$ is denoted by $4p$.

Except for the hydrogen atom, atoms are multi-electron systems and hence have a rather complex structure. Fortunately, the concept of single-electron functions still remains meaningful. The physical picture is that, focusing on a given electron, the effect of the remaining electrons, together with the Coulomb force from the nucleus, can be accounted for, to a good approximation, by a potential that is still spherically symmetrical. As was mentioned above, this has the consequence that the energy eigenstates (for single electrons) are also angular momentum eigenfunctions so that quantum numbers l and m can be assigned to them. In comparison with the hydrogen atom, the energy spectrum is, of course, modified. Nevertheless, the concept of a principal quantum number n can still be upheld, so that the above notation applies. It is important to note that the degeneracy known from the hydrogen levels is now lifted. This means that the energy levels for the single-electron states depend normally on *both* n and l.

6.2 Transitions

Interactions between systems often give rise to a temporal evolution that results in a change of the states of the systems. This means that the systems undergo transitions whose investigation is one of the main tasks of experimentalists. A rigorous theoretical treatment of such interaction processes through solution of the time-dependent Schrödinger equation (2.8) is often impossible. So we have to resort to approximative techniques known as perturbation theory. The mathematical treatment becomes rather simple when we specialize to short-time interaction processes. (A good example is scattering of a particle on another one. Then the particles affect one another appreciably only as long as they come close to one another.) In such circumstances, first-order perturbation theory suffices to give us a satisfactory result. Let me show this in some detail.

Let us consider two initially independent systems, labelled A and B, with Hamiltonian operators $\hat{H}^A$ and $\hat{H}^B$, respectively. The interaction is described by an interaction Hamiltonian $\hat{H}_{\text{int}}$, which naturally depends on observables of both systems A and B. So the Schrödinger equation (2.8) takes the form

$$i\hbar \frac{\partial |\psi\rangle}{\partial t} = (\hat{H}_0 + \hat{H}_{\text{int}}) |\psi\rangle, \quad \hat{H}_0 = \hat{H}^A + \hat{H}^B. \tag{6.15}$$

We suppose that the eigenstates and the eigenvalues of the unperturbed Hamiltonian $\hat{H}_0$ are known. (The eigenstates are the direct products of an eigenstate of $\hat{H}^A$ and an eigenstate of $\hat{H}^B$, and the eigenvalues are the sums of eigenvalues of $\hat{H}^A$ and $\hat{H}^B$.) As we know from the theory of linear operators, the eigenstates of a Hermitian operator give us an orthogonal basis in Hilbert space. So we can expand the wanted state $|\psi\rangle$ in terms of the eigenstates of $\hat{H}_0$, which we will simply denote by $|n\rangle$. The corresponding eigenvalues are E_n. It is advantageous to take explicitly into account the unperturbed evolution of the states $|n\rangle$ which means multiplying them by $\exp(-i\omega_n t)$, where $\omega_n = E_n/\hbar$. So we expand $|\psi\rangle$ in the form

$$|\psi\rangle = \sum_n c_n(t)e^{-i\omega_n t} |n\rangle. \tag{6.16}$$

Inserting (6.16) into (6.15) we obtain

$$i\hbar \sum_n \dot{c}_n(t)e^{-i\omega_n t} |n\rangle = \hat{H}_{\text{int}} \sum_n c_n(t)e^{-i\omega_n t} |n\rangle. \tag{6.17}$$

Forming the scalar product of this equation with an eigenstate $|f\rangle$, we benefit from the orthonormality of the eigenstates and arrive at the following system of first-order linear differential equations for the coefficients c_n

$$i\hbar \dot{c}_f(t) = \sum_n \langle f| \hat{H}_{\text{int}} |n\rangle \, e^{i(\omega_f - \omega_n)t} c_n(t). \tag{6.18}$$

It describes the evolution of the total system in the so-called interaction picture.

In first-order perturbation theory it is readily solved. We assume the total system to be initially, at $t = 0$, in just one state $|i\rangle$. So we have $c_n(0) = \delta_{ni}$, where δ_{ni} is the Kronecker symbol. We neglect the time dependence of the coefficients c_n on the right-hand side of (6.18), i.e., we replace them with the initial values. Then (6.18) is readily integrated to yield

$$c_f(t) = -\frac{1}{\hbar} \langle f| \hat{H}_{\text{int}} |i\rangle \, \frac{e^{i(\omega_f - \omega_i)t} - 1}{\omega_f - \omega_i}. \tag{6.19}$$

Hence the probability of finding the system in the state $|f\rangle$ at time t is given by

$$\left|c_f(t)\right|^2 = \frac{1}{\hbar^2}\left|\langle f|\hat{H}_{\text{int}}|i\rangle\right|^2 g(\omega_f - \omega_i, t),\tag{6.20}$$

where $g(\omega, t)$ stands for

$$g(\omega, t) = \sin^2\frac{\omega t}{2}\bigg/\left(\frac{\omega}{2}\right)^2.\tag{6.21}$$

This function has a pronounced peak of width $2\pi/t$ at $\omega = 0$. So it has the character of a resonance curve that describes conservation of the unperturbed energy. Actually, it allows the energy of the final state, E_f, to deviate from that of the initial state, E_i, by an amount of $\delta E_f \cong \pi\hbar/t$.

Now, in practical cases we are mostly dealing not with one particular final state but rather with a group of states with neighbouring energies. Then, a detector will not be able to discriminate between those states. So we have to sum the probability (6.20) over such a group of states whose energies lie around E_f, to get the observable detection probability $W(t)$. As usual, we will replace the summation over the group by an integration over the energy E_f, $\sum \ldots \rightarrow \int dE_f\sigma(E_f)\ldots$, where σ is the level density. Assuming that both the matrix element in (6.20) and σ are slowly varying functions of E_f within the domain of integration $E_i - \Delta E/2 \leq E_f \leq E_i + \Delta E/2$, we can take those functions outside the integral sign. The remaining integral, to a good approximation, equals $2\pi\hbar t$, when ΔEt is large compared with $2\pi\hbar$. Thus we arrive at the result

$$W(t) = \sum_f\left|c_f(t)\right|^2 = \frac{2\pi t}{\hbar}\left|\langle f|\hat{H}_{\text{int}}|i\rangle\right|^2\sigma(E_f),\tag{6.22}$$

where the so-called transition matrix element $\langle f|\hat{H}_{\text{int}}|i\rangle$ has to be taken at the resonance frequency $\omega_f = \omega_i$.

What makes the expression (6.22) especially attractive is the linear dependence on time. This allows us to define a time-independent transition probability per unit time, that is, a transition rate

$$w_{i\rightarrow f} = \frac{2\pi}{\hbar}\left|\langle f|\hat{H}_{\text{int}}|i\rangle\right|^2\sigma(E_f).\tag{6.23}$$

This is, in fact, a very important result that became famous as Fermi's Golden Rule. Applying it to specific interaction processes, we can predict quantitatively the outcomes of relevant measurements. What becomes obvious is that the strength of the effect is basically determined by the magnitude of the transition matrix element $\langle f|\hat{H}_{\text{int}}|i\rangle$ which is normally easy to calculate. In particular, a transition will be 'forbidden' when it vanishes.

Actually, the concept of (time-independent) transition rates traces back to Einstein, who as early as 1916 introduced it to describe electromagnetic emission and absorption processes, thus anticipating the idea that quantum processes are governed by statistical laws. It should be noted that this concept also underlies the exponential decay law (see Section 1.3).

Moreover, transition rates for various processes that take place simultaneously can be used to establish rate equations which provide valuable insight into their interplay. A good example is a laser system for which lasing conditions are readily derived from such equations.

Let us return to (6.23) and apply it, for illustration, to spontaneous emission. For a correct treatment of this process field quantization is indispensable. A simple procedure to describe a quantized electromagnetic field in free space is based on the mode concept. Modes form a complete set of field states into which a given field can be decomposed. In the present case it is natural to choose the modes as monochromatic polarized running plane waves, so they are characterized by a wave vector k and a polarization vector e (unit vector pointing in the direction of the electric field strength). For more details see Section 8.3.

So we have to specify the initial and the final state of the total system 'atom + radiation field' as follows. Initially, the atom, which we will idealize as a one-electron system, is in an excited state described by a wavefunction $\psi_b(x)$ and there are no photons present, i.e., all the field modes are in their vacuum states. When the transition has been completed, the atom is in a certain lower state $\psi_a(x)$ and just one field mode (k, e), which satisfies, to a good approximation, that the resonance condition (energy conservation law) is occupied by one photon, whereas all the other modes remain in their vacuum states. Note that the atomic wavefunctions are characterized by a full set of quantum numbers n, l, m, as we learned from the analysis of the hydrogen atom in Section 6.1.

In dipole approximation, the interaction Hamiltonian reads

$$\hat{H}_{\text{int}} = -\hat{d}\hat{E}, \tag{6.24}$$

where $\hat{d} = ex$ is the atomic electric dipole operator and $\hat{E}$ is the electric field strength operator taken at the centre of the atom. Since the interaction Hamiltonian factorizes, the transition matrix element does the same

$$\langle f | \hat{H}_{\text{int}} | i \rangle = \langle \psi_a | \hat{d} \cdot e | \psi_b \rangle \langle 1 | \hat{E} | 0 \rangle . \tag{6.25}$$

Here, we have put $\hat{E} = e\hat{E}$, and the photonic states $|0\rangle$ (vacuum state) and $|1\rangle$ (one-photon state) refer to a single mode (k, e) that becomes excited in the transition. The electric field strength operator $\hat{E}$ reduces to the contribution from the mode under consideration. It is a sum of two terms, which are proportional to the photon annihilation operator $\hat{a}$ and the photon creation operator $\hat{a}^\dagger$ for that mode, respectively. Only the latter operator comes into play in spontaneous emission. Actually, we have $\langle 1| \hat{a}^\dagger |0\rangle = 1$ (see Section 8.1).

Let us now specify the measuring device. For registration of the photon we will use a photodetector placed at some distance from the atom. It will detect such photons whose propagation directions fall into a solid angle $d\Omega$ that is determined by the sensitive surface of the detector and its distance. To get as much information as possible, we will place a polarizing filter in front of the detector so that we register only *e* polarized photons. This set-up does not enable us to resolve the frequency of the emitted light. However, this is just what we assumed in the derivation of (6.23)! So it remains only to calculate the energy density $\sigma(E_f)$. It is readily obtained from the mode density (8.18). The final result for the probability, per unit time, that a photon with polarization *e* is emitted into a solid angle $d\Omega$ is then found to be

$$ w = \frac{4\pi^2 \nu^3}{\hbar c^3} \left| \langle \psi_a | \hat{\boldsymbol{d}} \cdot \boldsymbol{e} | \psi_b \rangle \right|^2 d\Omega \tag{6.26} $$

(in electrostatic units), where ν is the mid-frequency of the emitted line.

So the existence of a nonvanishing transition matrix element for the electric dipole operator is a prerequisite of electric dipole radiation. Evidently, this depends sensitively on the quantum numbers l and m. Since the polarization vector *e* stands orthogonal to the propagation direction, the polarization dependence in (6.26) also implies a directional characteristic. Of great practical relevance is the frequency dependence of the emission rate (6.26). It predicts that spontaneous emission is strong in the ultraviolet and visible region of the spectrum, whereas it becomes weaker and weaker with growing wavelength.

It was already mentioned that in the present treatment of spontaneous emission no information on the linewidth of the emitted radiation, the so-called natural linewidth, can be gained. This is due to the short-time approach we used. In fact, the linewidth can be calculated only from a long-time solution of the Schrödinger equation. This was first done by Weisskopf and Wigner in 1930 (cf. Paul (2004)). A second remark has to do with the physical interpretation of our result. One might be tempted to consider the emission of a photon into a given solid angle as an 'objective' process that will take place, irrespective of whether a detector is present or not. Such an objectification is, however, erroneous. Had the emitted photon a well-defined emission direction in all experimental conditions, one could not understand how it 'interferes with itself' in a Young type interference experiment, for instance.

It should be noted that Fermi's Golden Rule is often applied to only one system *A* that is subjected to a disturbance. The latter can be due to external forces or to a second system *B* that is only slightly affected during the interaction so that the back action from the system A can be neglected. (The system B is often described classically.) Then the interaction Hamiltonian depends only on dynamical variables of the system under investigation. External influences are taken into account by a suitable potential or simply by given parameters. In some cases the interaction Hamiltonian becomes explicitly time-dependent. This must be observed in carrying out the time integration in (6.18).

6.3 Scattering

Scattering of particles on other particles or more complex systems yields valuable information on the underlying forces. Historically, Rutherford's pioneering experiments (from 1906 to 1913), in which α particles were scattered on atoms, were a milestone on the way to a realistic atom model. From the observation that the α particles were usually only slightly deviated – only in one of 8000 cases was there a strong deviation – it was concluded that an atom is virtually void, mass and electric charge being concentrated in point-like particles. This insight led to Rutherford's atomic model in which electrons revolve around a massive, positively charged centre like planets around the sun. In the following, a short introduction to quantum mechanical scattering theory will be given.

In a typical scattering experiment a beam of monoenergetic particles a is sent to a target consisting of particles b. When the particles come close to one another, they will interact, which results in a deviation of the incoming particles a: they will propagate, as outgoing particles, in various directions. What can be measured is the number of particles a that are scattered into a given solid angle. We suppose that the density of the particles in the incoming beam is so small that they do not interact with one another. Moreover, we focus on situations in which a particle is scattered by one target particle, at maximum. So what we observe is a manifold of elementary scattering processes in which only two particles are involved.

We describe the ingoing particle beam by a flux j that is defined as the product of the density of the particles ϱ and their velocity v, $j = \varrho v$. Then the number of particles that pass, per unit time, through a unit area perpendicular to the propagation direction is $j = \varrho v$. What is observed, on the other hand, is the number of particles dn that are scattered, per unit time, into a given solid angle $d\Omega$. The quantity dn will be proportional to both j and $d\Omega$ so that we can write

$$dn = \sigma(\vartheta, \varphi)j d\Omega, \tag{6.27}$$

where r, ϑ, φ are spherical coordinates with the direction of the incident beam as the polar axis. The quantity σ thus defined is known as the differential scattering cross-section. It has the dimension of an area and refers to an elementary scattering process in which one incoming particle interacts with one target particle. When the target is composed of N particles, we have to multiply σ by N. Integration of σ over the total solid angle yields the total scattering cross-section.

Let us first make an excursion to classical mechanics. A characteristic feature of scattering is that the two particles involved affect one another through short-range forces that depend on their relative coordinates only. Those forces have no influence on the centre-of-mass motion. Hence it is advantageous to separate the latter from the relative motion of the particles. This is done by using the centre-of-mass coordinate X and the relative coordinate x as new coordinates,

$$X = \frac{m_1 x_1 + m_2 x_2}{m_1 + m_2}, \quad x = x_1 - x_2, \tag{6.28}$$

where x_1, x_2 are the coordinates and m_1, m_2 the masses of the particles. The relations (6.28) are readily inverted to yield

$$x_1 = \frac{m_2}{m_1 + m_2} x + X, \quad x_2 = \frac{m_1}{m_1 + m_2} x + X. \tag{6.29}$$

Evidently, the description of scattering takes its simplest form in the centre-of-mass system in which the centre-of-mass is at rest $(X = 0)$. From now on we will use this reference frame. Then the two particles move in opposite directions; actually, their momenta $p_1 = m_1 \dot{x}_1, p_2 = m_2 \dot{x}_2$ add up to zero. It is advantageous to introduce a reduced momentum p of the system through the definition

$$p = \mu \dot{x} = \frac{m_2}{m_1 + m_2} p_1 - \frac{m_1}{m_1 + m_2} p_2, \quad \mu = \frac{m_1 m_2}{m_1 + m_2}, \tag{6.30}$$

where μ is the reduced mass. Moreover, it is easily seen that the kinetic energy of the system can be written as

$$T = \frac{p_1^2}{2m_1} + \frac{p_2^2}{2m_2} = \frac{p^2}{2\mu}. \tag{6.31}$$

This relation, together with (6.30), justifies the formal replacement of the two-particle system by one fictitious 'particle' of mass μ, position coordinate x and momentum p. Observing the relation $p_1 = -p_2$, we learn from (6.30) that p equals the momentum of particle 1, p_1, and the negative momentum of particle 2, $-p_2$. Since the (repulsive) forces between the actual particles can normally be derived from a potential $V(x)$, the 'particle' moves in a given potential field centred at $x = 0$. So the scattering problem takes a simple form, which is also well suited for a quantum mechanical analysis.

We describe the fictitious 'particle' by a wavefunction that has to fulfil the Schrödinger equation. There are two ways to solve the problem. The first is to treat scattering as a short-time interaction. Then we end up, in lowest-order approximation, with Fermi's Golden Rule (6.23). However, this approach has the drawback that it does not tell us anything about phase relations that might exist between waves scattered into different directions. Actually, we can get this information when we treat scattering as a stationary problem: a continuous flow of 'particles' is incoming and, as a result of scattering, there is a continuous flow of outgoing 'particles' which we will have to describe by just one wave that propagates in virtually all directions. So what we are looking for is a solution of the (time-independent) Schrödinger equation, which corresponds to a positive energy value E. Actually, when we consider elastic scattering, this energy is given by the kinetic energy of the 'particle', which is conserved in those conditions. So, unlike the situation in Section 6.1, where we determined energies as eigenvalues of the Schrödinger equation, the energy will now enter the latter as a given parameter.

Let us first write down the Schrödinger equation (6.1) in the form

$$(\Delta + k^2)\psi(x) = \frac{2\mu}{\hbar^2} V(x)\psi(x) \tag{6.32}$$

(also cf. (6.4)), where the abbreviation

$$k^2 = \frac{2\mu}{\hbar^2} E \tag{6.33}$$

has been introduced.

How to solve (6.32)? It turns out that a mathematical trick is helpful. (We follow here the analysis by Dawydow (1978).) Forget for a moment that the right-hand side depends on the wavefunction we are looking for, and treat (6.32) as an inhomogeneous differential equation with a given inhomogeneity. It is immediately seen that the general solution is a superposition of the general solution of the homogeneous equation and a special solution of the inhomogeneous equation. While it is easy to solve the homogeneous equation – which is well known from classical optics as the time-independent wave equation in free space – to find a special solution of the inhomogeneous equation is the real problem. Actually, this task can be reduced to the determination of the so-called Green's function $G(x, x')$ that in our case is required to satisfy the differential equation

$$(\Delta + k^2)G(x, x') = \delta(x - x'), \tag{6.34}$$

where x' is considered as a parameter. With the Green's function at hand, we obviously get a solution of the inhomogeneous differential equation on multiplying the inhomogeneity, written as a function of x', with $G(x, x')$ and integrating over x'.

In the present case the Green's function (of outgoing-wave type) is known to be

$$G(x, x') = -\frac{e^{ik|x - x'|}}{4\pi |x - x'|}, \tag{6.35}$$

i.e., it is a spherical wave emerging from a centre at $x = x'$. Since the special solution in question depends on the scattering potential, it will describe the scattered wave.

We still have to take into account the incoming 'particles'. Assuming them to have the momentum p_{in}, they are represented by a plane wave

$$\varphi_{in}(x) = e^{ik_{in}x}, \tag{6.36}$$

where $k_{in} = p_{in}/\hbar$ and $|k_{in}| = k$. This is a solution of the homogeneous equation. Hence we can simply add it to the special solution of the inhomogeneous equation. So we arrive at the following relation for the solution of (6.32):

$$\psi(x) = \varphi_{in}(x) - \frac{\mu}{2\pi\hbar^2} \int \frac{e^{ik|x-x'|}}{|x-x'|} V(x')\psi(x')d^3x'. \tag{6.37}$$

Though this is not an explicit solution, it is nevertheless very useful since it allows us to analyze the far-field behaviour of the solution, i.e., its approximate form at distances from the scattering centre $|x|$ that are large compared to the effective range of the scattering forces d. In fact, since the integration virtually extends over a small volume only in which the scattering potential $V(x')$ differs appreciably from zero, we can in the exponential approximate $k|x-x'|$ by $k|x| - k_{out}x'$, where $k_{out} = kx/|x|$ is the wave vector for a (scattered) particle that moves in the direction ϑ, φ. The denominator in (6.37), on the other hand, is a slowly varying function so that it can be replaced by $|x|$, to a good approximation. We thus arrive at the result

$$\psi(x) = \varphi_{in}(x) + A(\vartheta, \varphi)\frac{e^{ik|x|}}{|x|}, \quad |x| \gg d, \tag{6.38}$$

where

$$A(\vartheta, \varphi) = -\frac{\mu}{2\pi\hbar^2} \int e^{-ik_{out}x'} V(x')\psi(x')d^3x'. \tag{6.39}$$

So the scattered wave takes a form that meets our intuition: it is a spherical wave whose amplitude depends on the scattering direction. The squared modulus of the amplitude is the differential scattering cross-section defined above

$$\sigma = |A(\vartheta, \varphi)|^2. \tag{6.40}$$

What we still need, however, is an explicit solution for $\psi(x)$. Unfortunately, we are not able to find a rigorous solution of (6.37). So we have to resort to successive approximations. In first-order approximation we put $\psi(x) = \varphi_{in}(x) = \exp(ik_{in}x)$. This gives us the scattering amplitude in the so-called first Born approximation

$$A(\vartheta, \varphi) = -\frac{\mu}{2\pi\hbar^2} \int e^{i(k_{in} - k_{out})x'} V(x')d^3x'. \tag{6.41}$$

This means that, apart from a factor, the scattering amplitude is given by the Fourier component of the scattering potential that corresponds to the momentum transfer $p_{out} - p_{in} = \hbar(k_{out} - k_{in})$. We can express $A(\vartheta, \varphi)$ also as a transition matrix element

$$A(\vartheta, \varphi) = -\frac{\mu}{2\pi\hbar^2} \langle \varphi_{out} | V | \varphi_{in} \rangle, \tag{6.42}$$

where

$$\varphi_{out}(x) = e^{ik_{out}x} \tag{6.43}$$

and φ_{in} is given by (6.36). Inserting the result (6.42) into (6.40), we get just what we would have found in the transition rate approach. In fact, this should have been expected. What is really new is the representation (6.38) of the scattered *wave*, which also yields the phase of the scattering amplitude and allows us to study exchange effects that come into play when the two particles involved in the scattering process are indistinguishable. This problem will be discussed in some detail in Section 8.4.

Let me finally say a few words about inelastic scattering. Then at least one of the two interacting particles has an internal structure so that it may be excited in the scattering process. We consider the frequent situation where the target is formed of complex particles such as atoms or nuclei, whereas the projectile is an elementary particle. We suppose that the target particle is much heavier than the projectile. Then its position coincides with the centre of mass, to a very good approximation. We still use the centre-of-mass system for our description. So the target particle is always at rest, and only the motion of the projectile matters, whose position we will denote by x. In addition, we have to take into account the internal coordinates of the target particle ξ, whereby this symbol stands for a whole set of coordinates. These are, for instance, the position coordinates for the electrons in an atom. We will assume that the energy eigenvalues and the corresponding eigenfunctions for the target particle are known. The potential V that describes the interaction between the target particle and the projectile is now a function of x and ξ, $V = V(x, \xi)$. In electron-atom scattering, for instance, V describes two effects: (*i*) the Coulomb attraction of the incident electron due to the positively charged atomic nucleus and (*ii*) the Coulomb repulsion between the incoming electron and each of the bound electrons.

Inelastic scattering can be treated along the same lines as elastic scattering, the main task being the determination of the Green's function which is now a function of x, x' and ξ, ξ'. Let me omit the mathematical details, which you will find in textbooks (cf., e.g., Dawydow (1978)) and concentrate on the result. Let us assume that the target particle undergoes a transition from a lower energy level ϵ_i to a higher level ε_f, as a result of scattering. As in elastic scattering we describe the ingoing and the outgoing (scattered) particle by plane waves with wave vectors $\boldsymbol{k}_{in}$ and $\boldsymbol{k}_{out}$, respectively. However, their lengths k_{in} and k_{out} are no longer equal, since the particle becomes decelerated. It transfers part of its kinetic energy to the target particle, whereby the energy conservation law requires the relation

$$\frac{\hbar^2}{2\mu}k_{out}^2 = \frac{\hbar^2}{2\mu}k_{in}^2 + \varepsilon_i - \varepsilon_f. \tag{6.44}$$

to hold. When the right-hand side of this equation is positive, we speak of an open scattering channel. Otherwise the channel is closed. Then an excitation cannot take place, and the scattering will be elastic.

Now, with the above results (6.38) and (6.42) for elastic scattering at hand, it is easy to guess how we will have to describe inelastic scattering (again in the first Born approximation): focusing on those events in which the target particle has actually become excited, the general form of the wavefunction for the scattered particle will be retained, however, with two modifications: (*i*) the wave number $k = k_{out}$ is now determined by (6.44) and (*ii*) in the formula for the scattering amplitude (6.42) the transition matrix element refers to a transition of the whole system. So we expect the following relations to be valid. The wavefunction for the scattered projectile can be written as

$$\psi_{scattered}(\boldsymbol{x}) = A(\vartheta, \varphi)\frac{e^{ik_{out}|\boldsymbol{x}|}}{|\boldsymbol{x}|}, \quad |\boldsymbol{x}| \gg d, \tag{6.45}$$

where

$$A(\vartheta, \varphi) = -\frac{\mu}{2\pi\hbar^2}\langle\varepsilon_f, \boldsymbol{k}_{out}| V(\boldsymbol{x}, \boldsymbol{\xi}) |\varepsilon_i, \boldsymbol{k}_{in}\rangle. \tag{6.46}$$

This is, in fact, what a thorough analysis yields.

It should be noted that the relationship (6.40) between the scattering amplitude and the differential scattering cross-section has to be modified too. Since the scattered particles move with a velocity that is smaller by a factor of k_{out}/k_{in} than that of the incident particles, the differential scattering cross-section becomes reduced by the same factor

$$\sigma = \frac{k_{out}}{k_{in}}|A(\vartheta, \varphi)|^2. \tag{6.47}$$

6.4 Tunnelling effect

What happens when a particle is running against a barrier described by a potential 'hill'? Let us consider the problem as one-dimensional, for simplicity. We suppose that the potential $V(x)$ grows from zero to a maximum, from where it falls to zero again. Then a particle coming from, say, the left, will become decelerated when it penetrates into the barrier, since it experiences a repulsive force $f = -dV(x)/dx$. In classical mechanics, the particle's velocity v at a given position x is determined by the energy conservation law

$$T = \frac{m}{2}v^2 = E - V(x). \tag{6.48}$$

Here, m is the mass of the particle, T its kinetic energy at position x and E the total energy of the system (the kinetic energy at a large distance from the barrier), which is conserved. Let us now assume that E is lower than the maximum of the potential. Then the particle cannot pass through the barrier since, according to (6.48), its kinetic energy would become negative when it approaches the potential maximum. This, however, is impossible. So the particle will penetrate into the barrier until it reaches a turning point x_{turn} defined through the relation $E = V(x_{turn})$, where its motion becomes reversed.

Quantum theory, however, tells us that this simple classical picture must be revised. It predicts that the particle, after having reached the turning point, will penetrate, with some probability, a bit deeper. This even gives the particle a chance of passing through the barrier, when the latter is thin. This phenomenon became known as the tunnelling effect. It turned out that it is indeed of practical relevance. Some examples will be given below.

First, let me briefly explain the quantum mechanical description of tunnelling. To make the analysis as simple as possible, we consider a potential of rectangular form. As in scattering theory, we will look for stationary solutions of the (time-independent) Schrödinger equation. We suppose a constant flux of particles with energy E to come from the left. Most of the particles will be reflected, which gives rise to a counter-propagating flux. As a result of tunnelling, there will be a (rather weak) outgoing flux of particles too. All these three fluxes are readily described quantum mechanically. Since the particles are propagating freely, the fluxes outside the barrier are represented by running plane waves with wave number $k = \sqrt{2mE/\hbar^2}$. More interesting is the question of how to describe the particles when they are inside the barrier. Making again the ansatz of a travelling wave $\exp(ik'x)$, the Schrödinger equation gives us $k'^2 = 2m(E - V)/\hbar^2$. Hence k'^2 is negative. This is just the counterpart of the classical result that the kinetic energy is negative. However, contrary to the classical description, this will not irritate us; we simply conclude that k' must be imaginary, $k' = \pm i\kappa$, where $\kappa = \sqrt{2m(V - E)/\hbar^2}$ is positive. Thus the wavefunction no longer shows oscillations; it is simply an exponential, $\exp(\kappa x)$ or $\exp(-\kappa x)$.

Now the tunnelling problem is easily solved. We have to determine the amplitudes of the waves mentioned above such that the matching condition is fulfilled at the two boundaries of the barrier: both the wavefunction and its first derivative are required to be steady there. In this way, the probability for a particle to pass through the barrier is found to be

$$w = \frac{1}{1 + \frac{1}{4}(\frac{k}{\kappa} + \frac{\kappa}{k})^2 \sinh^2 \kappa d},$$ (6.49)

where d is the thickness of the barrier.

From this result it becomes obvious that tunnelling takes place more frequently the thinner the barrier is. On the other hand, when the barrier is rather thick ($\kappa d \geq 1$) and k and κ are of the same order of magnitude, the probability (6.49) can be approximated by

$$w \approx 4e^{-2\kappa d}.$$ (6.50)

It should be noticed that the tunnelling effect also comes into play when a particle is surrounded with a potential barrier that is not too thick. Then the particle is in a bound state which, however, is unstable.

It becomes obvious from the above analysis that it is the wave character of particles that makes tunnelling possible. Actually, an analogue of tunnelling is known from classical optics. It can be observed in connection with total reflection. Let me first explain the latter phenomenon. We consider a light wave that is incident from an optical medium with index of refraction n_1 on a second medium with a smaller index of refraction $n_2 < n_1$, for instance air. Then, provided that the angle of incidence exceeds a critical value, no refracted wave can build up in the second medium, so that the electromagnetic energy carried by the incident wave is fully transferred to the reflected wave, i.e., we have total reflection. A closer inspection shows that the electromagnetic field nevertheless penetrates a bit into the second medium. This is, however, an unusual light wave: it propagates along the boundary, its amplitude being strongly damped in the transverse direction. So it is an inhomogeneous wave that has been termed an evanescent wave. When the second medium is extremely thin (of the order of the wavelength of the light wave) and a third medium with a higher index of refraction (usually it is made of the same material as the first medium) is brought into contact with it, electromagnetic energy is coupled into the third medium, which results in the excitation of a (weak) propagating wave of the type we are familiar with. (One then speaks of frustrated total reflection.) So there is a perfect analogy with the quantum mechanical tunnelling effect.

The first to come up with the idea of quantum tunnelling was George Gamow. In 1928 he recognized that this effect could explain α decay, a phenomenon known from natural radioactivity. Some chemical elements had been observed to emit strongly ionizing radiation that had been shown to consist of ^{4}He nuclei.

Gamow's picture was that an α particle, as a stable configuration of two protons and two neutrons, pre-exists in the nucleus. The strong attractive nuclear force that dominates over Coulomb repulsion hinders the α particle from escaping. This situation can be modelled by a potential well, as a function of the distance r between the residual nucleus and the α particle. Since the nuclear force has an extremely short range, the potential rises very steeply at the edge of the well. It reaches a maximum at a positive energy, from where it falls gradually to zero with growing r. This decrease describes the Coulomb repulsion (note that both the residual nucleus and the α particle are positively charged), which is now present alone. So a potential barrier is formed. Provided that the α particle has a positive energy when it is still in the nucleus, it may tunnel through this so-called Coulomb barrier. Gamow's model explained the statistical character of the decay (which becomes manifest in the exponential decay law). Moreover, it accounted for the observed dependence of the transition rates on the energy released in the decay.

Similarly, the Coulomb barrier might be overcome from outside by tunnelling. This effect is, in fact, very important in nuclear physics. It opens the possibility of inducing nuclear reactions with positively charged particles that have moderate (kinetic) energies, compared with the maximum of the Coulomb barrier. In particular, tunnelling is crucial to thermonuclear fusion, where thermally excited light nuclei react to form heavier nuclei. In these reactions, energy is gained (in the form of kinetic energy of the newly produced nuclei) that stems from the mass defect (the sum of the masses of the particles is greater before the reaction than afterwards), according to Einstein's famous formula $E = mc^2$ that establishes an equivalence of mass m and energy E. Owing to the large binding energy of ^{4}He nuclei, the energy output is especially high when those nuclei are produced.

Actually, the 'burning' of hydrogen to ^{4}He – either directly, via the production of deuterium (^{2}H) and ^{3}He as intermediate steps, or in a more complex cycle in which ^{12}C is consumed and, however, finally recovered – is the fundamental mechanism that heats up stars (including our sun!) and thus makes them shine. The concept of producing ^{4}He from light elements such as deuterium, tritium and lithium, was realized in the thermonuclear, or hydrogen, bomb, and it underlies present attempts to construct a thermonuclear fusion reactor. The point is that all those fusion processes are induced by thermally excited nuclei. Hower, even at temperatures of about 10^8 K, as they exist in the sun, the mean thermal energy is distinctly lower than the maximum of the Coulomb barrier. So tunnelling is indispensable, and its importance for cosmology can hardly be overestimated.

Quantum tunnelling is not confined to such extraordinary conditions. There are also many technical applications on a microscopic scale. For instance, the above-mentioned α decay has an analogue in field emission from a metallic surface.

In the absence of an external electric field, electrons are prevented from leaving the metal by a potential step. (A certain work, the so-called work function, must be done to release an electron.) However, when a strong static electric field (of proper polarity) is applied (in practice, this is achieved with the help of a fine metallic tip that is placed close to the metallic surface), the potential falls steeply with growing distance from the surface. So an electron 'sees' a potential barrier which it might tunnel through. Thus electrons are continuously emitted.

Another example is the tunnel, or Esaki, diode. This semiconductor diode has a heavily doped *p–n* junction. With increasing voltage applied in forward bias operation, the electron states in the conduction band on the *n* side become aligned with the holes in the valence band on the *p* side. So electrons can tunnel through the junction barrier, thus producing a current. However, as the voltage increases further, the mentioned states become misaligned, resulting in a decreasing current. This corresponds to a negative resistance that is a characteristic feature of the tunnel diode.

Quite generally, tunnelling plays an important role in semiconductor and superconductor physics. It should be noted, however, that this process is not always beneficial. It may become a source of current leakage.

Last but not least, tunnelling has found an impressive application in microscopy. In the scanning tunnelling microscope, it is utilized to image a surface of a conducting material, with a resolution that may reach, under favourable conditions, 10^{-2}Å in the vertical and 1Å in the lateral direction. So individual atoms become 'visible'! The measuring principle is as follows. A voltage is applied between the surface and an atomically sharp metallic tip that is placed close enough to the surface so that a tunnelling current will flow from the surface to the tip or in the reverse direction, depending on the polarity. Now, the tip is moved over the surface in such a way that the tunnelling current remains constant. This is achieved using a feedback loop that adjusts, with the help of a piezoelectric element, the distance between the surface and the tip. So, scanning the tip over the surface, it will follow the surface profile. The voltage applied to the piezoelectric element is used as a measuring signal from which an image of the surface can be constructed.

7

Conservation laws

As in classical physics, also in quantum mechanics: physical processes are bound to obey conservation laws. However, there is an important difference. While in classical physics we can suppose that all variables have definite values (irrespective of whether we know them or not), quantum mechanics allows them to be intrinsically uncertain. In those cases conservation laws apply only to ensemble averages (expectation values), and hence are less informative. They show their full strength in measurement processes. Then the measured values satisfy the respective conservation laws in any individual experiment. Let me remind you of spin measurements on a system of two spin-$\frac{1}{2}$ particles, being in a state with zero total spin, which we used in Section 5.5 to illustrate the EPR paradox. The spin components, with respect to an arbitrarily chosen direction, of the two particles are indefinite; however, measuring them yields definite values ($+\frac{1}{2}$ or $-\frac{1}{2}$) that add up to zero, thus confirming the angular conservation law.

In the following, we will consider some realistic experiments for which conservation laws provide a simple explanation.

7.1 Energy and momentum conservation

For an understanding of the interaction between matter and radiation, Einstein's photon concept (actually, he spoke of 'light quanta') proved to be very helpful. He assumed photons to be particle-like constituents of electromagnetic radiation, in particular light. They have the energy $h\nu$, where ν is the frequency of the radiation field. (Since the electromagnetic spectrum extends over an extremely wide range, there is a huge number of different sorts of photons.) The crucial point is that photons are generated or annihilated only as a whole. This property gives rise to typical quantum effects.

Einstein himself, being led to the 'light quanta hypothesis' by thermodynamic considerations (Einstein, 1905), used the photon concept to give a simple explanation of the photoelectric effect, which is nothing but an energy balance: the energy of an incident photon is partly consumed to release an electron from the atomic structure, and the residual energy is converted into kinetic energy of the electron:

$$hv = A + \frac{m}{2}v^2, \tag{7.1}$$

where A is the work function, m the electron's mass and v its velocity. Actually, the predicted frequency dependence of the kinetic energy was confirmed no earlier than 1916 through careful measurements by Millikan. Last but not least, (7.1) gives us a convenient means of measuring Planck's constant h.

Similarly, Bohr's second postulate (1.3) is readily interpreted as an energy balance. The energy lost (gained) by an atom in a transition equals the energy of the emitted (absorbed) photon. But why should only one photon be involved in an atomic transistion? In fact, two- (even many-) photon processes are possible. The transition probability is, however, extremely small. In the absorption case it is proportional to the nth power of the light intensity (n number of simultaneously absorbed photons). At least two-photon absorption becomes feasible when the atoms, or molecules, are exposed to the high-intensity radiation from lasers.

Let us focus on two-photon absorption, which is of practical importance, for instance in high-resolution laser spectroscopy in gases. Assuming the light field to oscillate at two frequencies, v_1 and v_2, Bohr's frequency condition (1.3) has to be generalized in the form

$$E_m - E_n = h(v_1 + v_2), \quad E_m > E_n. \tag{7.2}$$

Actually, photons possess not only a given energy, but also a definite momentum $\hbar k$, where k is the wave vector. Strictly speaking, this momentum, as a vector, is defined only in special conditions: either the photon belongs to a (quasimono-chromatic) field that propagates in a given direction, or the photon's propagation direction can be reconstructed from a measurement of its position (with the help of a photodetector) and the knowledge where it started from.

The existence of photonic momentum was first confirmed in Compton scattering. This phenomenon was discovered by A. H. Compton in 1923, when he directed an X-ray beam of well-defined frequency and direction at a paraffin block. He observed the scattered radiation and found that its wavelength varied with the scattering angle ϑ in a strange way: the wavelength difference, with respect to the wavelength of the incident radiation, followed the law

$$\Delta\lambda = \Lambda(1 - \cos\vartheta). \tag{7.3}$$

Here, the constant Λ, the so-called Compton wavelength, is independent of the wavelength of the incident radiation, which excludes an interpretation of the formula as a Doppler effect.

Compton and, independently, Debye could explain the effect as scattering of X-ray quanta off free electrons. (These are the electrons present in carbon atoms. Since they are slightly bound, they can be considered as approximately free.) They showed that the energy and the momentum conservation law actually suffice to deduce the result (7.3). An important point was that they used the full quantum mechanical picture of photons, i.e., they ascribed both energy and momentum to a photon.

Let me go into some detail. In the initial state the electron is at rest and hence has zero momentum, and its energy is (in relativistic description) the rest energy $m_0 c^2$ (m_0 electron mass). In addition, an incident photon with energy $h\nu_0$ and momentum $\hbar k_0$ is present. After the interaction the electron moves with velocity v. So it has the momentum mv and the energy mc^2, where $m = m_0 / \sqrt{1 - v^2/c^2}$. Furthermore, there exists a scattered photon with energy $h\nu$ and momentum $\hbar k$. Then the energy and momentum conservation laws are readily written down:

$$\frac{m_0 c^2}{\sqrt{1 - v^2/c^2}} + h\nu = m_0 c^2 + h\nu_0, \tag{7.4}$$

$$\frac{m_0 v}{\sqrt{1 - v^2/c^2}} + \hbar k = \hbar k_0. \tag{7.5}$$

With a little algebra, from these equations Compton's formula (7.3) is indeed recovered. Moreover, the calculation yields the following expression for the Compton wavelength:

$$\Lambda = \frac{h}{m_0 c}, \tag{7.6}$$

which shows that it has, in fact, universal character.

Of special interest is the fact that individual scattering events can be observed in a cloud chamber, as was done by Compton and Simon in 1925. The propagation direction of the incident photon is fixed through the experimental set-up. The direction of motion of the electron hit by the photon can be observed from the first part of the track it produces, and its kinetic energy can be determined from the total length of the track. The propagation direction of the scattered photon, on the other hand, is indicated by the starting point of an electron track that originates from a second Compton process in which the photon becomes newly scattered. In this way individual scattering processes can actually be observed in their details. Moreover, in 1925 Bothe and Geiger provided evidence to support the assumption inherent in the theoretical analysis, that ejection of an electron and emission of a scattered photon occur simultaneously. They measured the corresponding temporal correlations by counting coincidences between the signals from two Geiger counters that responded to the electron and the photon, respectively. In fact, they found more coincidences than could be expected from random events. So it appears justified to rely on the validity of conservation laws even in individual cases.

In particular, Pauli's belief in the conservation laws was so strong that in 1930 he postulated the existence of a new particle to solve the puzzle of beta decay. This phenomenon was known from natural radioactivity: a mother nucleus emits an electron, whereby it becomes transformed into a daughter nucleus that has one proton more and one neutron less. So the production of the electron is associated with the conversion of a neutron into a proton. Measurements of the energy of the emitted electron showed an amazing result. This energy was not sharp, as was to be expected from energy conservation, but was distributed over an interval from zero to a maximum energy, the distribution function having a maximum at somewhat less than half the maximum energy. This was, in fact, in distinct contradiction to the energy conservation law: the mass difference between the mother and the daughter nucleus defines, through Einstein's energy–mass relation, the released energy, which therefore has a fixed value that coincides with the maximum electron energy. However, due to the large nuclear mass compared to the electronic one, all this energy should have been imparted to the electron in any individual case. So it was a matter of fact that energy was definitely missing. (Using calorimetric methods, it was observed on a sample that only about 40 per cent of the energy corresponding to the mass difference were appearing as beta ray energy.) So where had the missing energy gone? Furthermore, the angular conservation law is violated since the electron has spin $\frac{1}{2}$, whereas the spins of the nuclear states connected by beta decay differ by either zero or an integral value.

Those peculiarities led Pauli to the conclusion that a second particle must be emitted, together with the electron, that carries away the missing energy. It must be neutral and have spin $\frac{1}{2}$ to ensure momentum conservation. Such a mysterious particle – it was baptized neutrino ('little neutron') by Fermi – had never been observed. Its average penetration depth in lead was calculated by Bethe and Peierls to be one thousandth of a light year, really a fantastic figure. So there was no hope that this ghost-like particle could ever be observed. However, 25 years later it turned out that this pessimistic view had to be given up. The availability of nuclear reactors that produced intense neutrino beams together with the construction of detectors that consisted of a huge number of particles (tons of appropriate substances were employed!) made it possible to detect nuclear reactions that were triggered by neutrinos. So Pauli's prediction proved to be a triumph of theoretical reasoning.

7.2 Angular momentum conservation

Conservation of angular momentum plays a crucial role in the interaction of electromagnetic radiation with matter. The reason is that the emission, or likewise the absorption, of photons goes hand in hand with transitions between atomic states which have not only fixed energies but also definite angular momenta l_a (initial state) and l_b (final state). So, to ensure momentum conservation, the emitted photon must carry off a certain amount of angular momentum, or the absorbed photon must impart it to the atom. In fact, photons can be in different states with angular momentum $L = 1, 2, \ldots$ (corresponding to dipole, quadrupole, ... radiation).

In Section 4.2 we already specified quite generally the transitions that are allowed by angular conservation. In the following, we will focus on optical transitions in atoms, which are dipole transitions, as a rule. (Compared with them, quadrupole transitions are extremely weak.) So the photons have angular momentum $L = 1$, and hence angular momentum conservation requires that the initial and the final atomic state differ in their angular momenta by unity: $\Delta l = l_b - l_a = \pm 1$. (The electronic spin state does not change in the transition. So it can be disregarded.) This restriction is known as a selection rule. Moreover, the angular momentum components (magnetic quantum numbers) $m_b = -l_b, -l_b + 1, \ldots, l_b$ and $m_a = -l_a, -l_a + 1, \ldots, l_a$ are bound to fulfil the selection rule $\Delta m = m_b - m_a = \pm 1, 0$.

As was explained in Section 6.2, the transition probability for emission of a photon polarized in the e direction is proportional to the squared modulus of the product of the transition matrix element for the electric dipole operator, $\langle \psi_b | e x | \psi_a \rangle$, and the unit vector e. So from the knowledge of the transition matrix elements we can get information on the polarization properties of the emitted radiation. We need only know which transition matrix elements are different from zero. To answer this question, it is sufficient to integrate over the angles ϑ, φ. Hence only the spherical harmonics (the angular momentum eigenfunctions, see (6.9)) come into play, and from their properties it follows that the nonvanishing transition matrix elements are:

$$\Delta m = 0 \text{ transition:} \quad \langle \psi_b | z | \psi_a \rangle = \langle \psi_b | r \cos \vartheta | \psi_a \rangle, \tag{7.7}$$

$$\Delta m = +1 \text{ transition:} \langle \psi_b | x + iy | \psi_a \rangle = \langle \psi_b | r e^{i\varphi} | \psi_a \rangle, \tag{7.8}$$

$$\Delta m = -1 \text{ transition:} \langle \psi_b | x - iy | \psi_a \rangle = \langle \psi_b | r e^{-i\varphi} | \psi_a \rangle. \tag{7.9}$$

It is important to note that the degeneracy of the atomic states can actually be lifted by applying a static magnetic field (whose direction defines now the quantization axis z physically). Then the energy levels split according to their magnetic quantum numbers (Zeeman effect) so that various emission lines can be observed that differ in their polarization properties. From the above values for the transition matrix elements we can readily draw the following conclusions. (*i*) When the radiation emitted orthogonally to the magnetic field (say, in the y direction) is examined, one finds the lines to be linearly polarized, namely z polarized for $\Delta m = 0$ and x polarized for $\Delta m = \pm 1$. (*ii*) The radiation emitted along the field direction, on the other hand, will be left- or right-handed circularly polarized, depending on whether Δm equals $+1$ or -1. Actually, this follows immediately from the fact that the photon has the angular momentum component $m = -1$ in the first case and $m = +1$ in the second.

It should be emphasized that the transition matrix elements in question also describe the absorption process. In fact, the transition rates for absorption and (spontaneous) emission differ only by a factor that characterizes the intensity of the incident radiation. So, in the absence of Zeeman splitting, one can selectively populate sublevels in the excited state: identifying the quantization axis with the direction in which radiation is sent to the atoms, circularly polarized light will induce only $\Delta m = +1$ or $\Delta m = -1$ transitions, depending on whether the light is right- or left-handed circularly polarized. So only certain sublevels will be 'pumped', and the others not. This results – in the interplay with spontaneous processes that tend to deplete excited levels – in a nonuniform occupation of the sublevels in the excited state. This technique, introduced by A. Kastler in 1950, became known as optical pumping.

8

Spin and statistics

8.1 Photons and electrons

One of the great technical achievements of the second half of the 20th century was the invention of the laser. With its help extremely intense light could be produced. Furthermore, it even became posssible to excite a single resonator mode so that both the frequency and the propagation direction of the emitted radiation field are extraordinarily sharp. The mere existence of such high-intensity single-mode fields – the real problem that was ingeniously solved was to generate them – is no matter of surprise. We know from classical electromagnetic theory that the amplitude of a monochromatic electromagnetic field can take on arbitrarily large values. This is a direct consequence of the linearity of Maxwell's equations. What does this mean for quantized radiation fields? Certainly, the mentioned property of classical fields must also be retained in the quantum mechanical description. Since the electromagnetic energy is quantized in 'packets' of magnitude $h\nu$ (photons), the number of photons in a given mode must be unlimited, in principle.

All this is very well known. However, it becomes noteworthy when we compare the behaviour of photons with that of electrons. Amazingly, it turns out that an electronic quantum state can be occupied by *one* electron, at maximum. (Strictly speaking, in the description of the electronic state the spin variable must be included.) This drastic difference in the collective behaviour of photons and electrons could, in fact, not be expected from what we know from single photons and electrons. Both of them have particle-like as well as wave-like properties. However, when many of them are present, they exhibit quite different a demeanour. Photons are very social, they like intimacy, whereas electrons are 'lone wolves', unable to tolerate the presence of even one companion in their 'district'. So you can never achieve laser-like action with electrons!

This curious behaviour of electrons is demonstrated beyond all doubt by the structure of atoms. It is well understood that the periodic system of elements established by the chemists D. I. Mendeleev and, independently, L. Meyer, can be founded on Pauli's exclusion principle saying that any atomic single-electron energy level (including the spin state) cannot be occupied by more than one electron. (Those electronic quantum states can approximately be defined in an n-electron atom by describing the residual $n - 1$ electrons, jointly with the Coulomb attraction from the nucleus, by an effective potential. They are multiplied by states that describe the spin orientation.) So with increasing ordinal number in the periodic system, which equals the charge number (the number of protons and, hence, also that of the electrons), the possible energy levels will be 'filled' one after the other, starting from the lowest energy level and proceeding to the higher ones, by just two electrons, when we disregard the electronic spin. (Strictly speaking, this applies to the atomic ground state.) This underlines the enormous impact the 'anti-social conduct' of the electrons has on what our (macroscopic!) world looks like.

Formally, the above-mentioned collective properties of photons, on the one hand, and electrons, on the other hand, can be reproduced by an appropriate choice of commutation rules. Let us first consider photons. The usual way to quantize a radiation field is to replace the (complex) classical amplitude of a field mode, up to a normalization factor, by a photon annihilation operator $\hat{a}$, and, consequently, the complex conjugate of the amplitude by a photon creation operator $\hat{a}^\dagger$ (the Hermitian conjugate of $\hat{a}$). Those operators are subjected to the commutation relation

$$\hat{a}\hat{a}^\dagger - \hat{a}^\dagger\hat{a} \equiv \left[\hat{a}, \hat{a}^\dagger\right] = \mathbf{1}. \tag{8.1}$$

It is not difficult to show that the operator $\hat{n} = \hat{a}^\dagger\hat{a}$ has the eigenvalues $0, 1, 2, \ldots$ and hence represents the number of photons. Applying the photon creation and annihilation operators to an n-photon state, we get the results

$$\hat{a}^\dagger |n\rangle = \sqrt{n + 1} \, |n + 1\rangle, \tag{8.2}$$

$$\hat{a} |n\rangle = \sqrt{n} \, |n - 1\rangle, \tag{8.3}$$

which justify the names of the operators.

By the way, the relation (8.2) is of fundamental importance for laser action. It describes stimulated emission into the mode already excited. When the laser field has been built up to some extent, excited atoms will preferably emit into the mode already excited, following the biblical text (Matthew 25:29) saying that 'He who has will be given more.'

Electrons, on the other hand, are (in the so-called second quantization) characterized by annihilation and creation operators (for one mode), $\hat{b}$ and $\hat{b}^\dagger$, that satisfy the *anticommutation* relation

$$\hat{b}\hat{b}^\dagger + \hat{b}^\dagger\hat{b} \equiv \left\{\hat{b}, \hat{b}^\dagger\right\} = 1, \quad \left\{\hat{b}, \hat{b}\right\} = \left\{\hat{b}^\dagger, \hat{b}^\dagger\right\} = 0, \tag{8.4}$$

from which it follows that the electron number operator (for a given mode) $\hat{b}^\dagger \hat{b}$ can take on the eigenvalues 0 and 1 only. Application of the creation and annihilation operators to number states yields

$$\hat{b}^\dagger |0\rangle = |1\rangle, \quad \hat{b}^\dagger |1\rangle = 0, \tag{8.5}$$

$$\hat{b} |0\rangle = 0, \quad \hat{b} |1\rangle = |0\rangle. \tag{8.6}$$

But what is the *physical* reason for the totally different behaviour of photons and electrons? The answer is: it is not the difference in the electric charge but rather in the spin which equals 1 (in units of $\hbar$) for photons and $\frac{1}{2}$ for electrons. Actually, it turned out that what matters are not the specific spin values mentioned but the fact that the spin is an integer in one case and a half-integer in the other. Based on experimental evidence, the following statement could be made: all particles with integer spin obey Bose (or Bose–Einstein) statistics, and hence they were called bosons, whereas all particles with half-integer spin obey Fermi (or Fermi–Dirac) statistics, and hence are referred to as fermions. I will explain these two types of nonclassical statistics in some detail in Section 8.3. At the moment, I would like to stress only that a basic feature of Bose statistics is that any quantum state can be occupied by many bosons, whereas Fermi statistics allows only a single fermion to occupy a quantum state. The mentioned interconnection between spin and statistics is really amazing. It is easily formulated but very hard to prove. (In 1940 Pauli gave an intricate proof which, I believe, only few theorists read.)

Noticing that there exist isotopes of chemical elements that differ only in that one nucleus has a neutron (a spin-$\frac{1}{2}$ particle) more, we must put up with the almost incredible fact that those isotopes, since their spin is integer in one case and half-integer in the other, differ drastically in their statistical behaviour. To me, it looks like a miracle that just one additional neutron should have so radical an effect. So it remains only to cite Newton, who wrote in his *Opticks*: 'There is no argument against facts and arguments.' And facts are in the present case provided by the so-called Bose–Einstein condensation which could be observed only on atoms with integer spins. The experimental verification of this phenomenon predicted by Einstein in 1925 – at extremely low temperatures, a large fraction of the particles 'condenses' into the lowest energy state – is one of the highlights of physical research in the last two decades. So let me briefly explain it.

8.2 Bose–Einstein condensation

The key to the generation of Bose–Einstein condensates is extremely strong cooling. Indeed, what has to be achieved experimentally is that the quantum mechanical wavepackets of the individual particles overlap, which clearly makes them indistinguishable. In these conditions a gas will undergo a phase transition, forming a strongly localized cloud of atoms that are all in the same state. The spatial extent of the wavepackets is the position uncertainty associated, through Heisenberg's uncertainty relation, with the thermal momentum distribution, which, therefore, must be made exceptionally narrow. This requires penetration into a temperature domain never reached before. Actually, temperatures of the order of about 10 μK had to be realized experimentally. The very effective optical (or laser) cooling technique explained in Section 2.6.1 meets with its limits at temperatures of the order of 100 μK. So a new idea was in demand.

The solution to the cooling problem was rather simple, at least in principle. If you have an ensemble of atoms that are strongly cooled down already, remove atoms with great velocities. This will diminish the mean value of the kinetic energy of the residual gas, and hence reduce its temperature. (Elastic two-particle collisions between the atoms will give rise to rethermalization.) This principle is in fact well known as evaporative cooling. The researchers hunting for Bose–Einstein condensates devised a clever technique which made evaporation an *induced* process. In a first step, they transferred laser-cooled atoms into a magnetic trap. This apparatus consists of a static magnetic quadrupole field that is produced by two coils with opposing currents. Now, in an inhomogeneous magnetic field a magnetic dipole experiences a force whose direction depends on the orientation of the dipole. In a magnetic quadrupole field this force is everywhere attractive (except the centre of the field, where the force vanishes) for a suitably oriented dipole. So this device acts as a trap, when atoms are excited in a properly chosen hyperfine level, which leads to the desired orientation of their nuclear spins. Now, when the spin flips (the spin component reverses its sign), the force becomes repulsive and the particle gets repelled from the trap. This is just an evaporation process!

A spin flip is readily produced by irradiating the atom with a microwave field resonant with the transition associated with the spin flip. The resonance condition provides us, in fact, with an opportunity to *selectively* induce a spin flip. To this end, we utilize the Doppler effect, which makes an atom 'see' the Doppler shifted frequency $v' = v(1 + v/c)$ when it is moving with velocity v opposite to the propagation direction of the radiation field. Via the resonance condition $v' = v_0$, where v_0 is the transition freqency, we can select a given velocity group of atoms by properly tuning the radiation frequency. This is the principle on which induced evaporative cooling rests. In this way, it became possible to reduce the temperature by a factor of about 10.

A serious obstacle on the road to Bose–Einstein condensation is the well-known fact that gases usually undergo phase transitions to the liquid or the solid state when they are cooled down. So care has to be taken that this does not happen with the gas chosen as a candidate for Bose–Einstein condensation. The only way to prevent ordinary condensation is to work at very low densities. The criterion is that three-body collisions that are needed for the formation of molecules or clusters are very rare so that binary collisions take place most frequently. Actually, this condition requires extremely low densities of about 10^{14}cm^{-3}. Hence a Bose–Einstein condensate can only be formed from a rather small number of atoms (up to 10^7 in the case of alkali atoms).

Using the described cooling technique and observing the density limitation, researchers succeeded, in fact, in producing Bose–Einstein condensates. The first to do this, in 1995, were the group of E. Cornell and C. Wieman at Boulder (Anderson *et al.*, 1995) and, independently, the group of W. Ketterle at MIT (Davis *et al.*, 1995). In 2001 the mentioned three persons were jointly awarded the Nobel prize for their spectacular achievements. Absorption imaging revealed that, after passing the transition temperature, the atoms were indeed confined to an extremely small volume with dimensions of about $100\ \mu\text{m}$.

A further highlight was the demonstration, by Ketterle and his coworkers, that two parts into which a Bose–Einstein condensate had been divided could be made to interfere. Actually, interference fringes were observed.

Finally, I would like to mention that vortices can be produced in Bose–Einstein condensates by spinning laser beams around the condensate. Usually, the vortices whose diameters are less than $1\ \mu\text{m}$ show up in the form of a regular array. A vortex is a collective motion of atoms similar to water swirls. The atoms undergo circular motions around the vortex axis such that the angular momentum is the same for all of them. Quantum mechanics requires atomic angular momenta to be quantized, restricting the allowed values to multiples of $\hbar$. In the observed tiny vortices it was just $\hbar$. Since the occurrence of quantized vortices is a characteristic property of superfluids, there is an interesting similarity between Bose–Einstein condensation and superfluidity.

8.3 Statistics

A frequent physical situation is that a system of many identical particles is in thermodynamic equilibrium at a temperature T. Usually a great number of one-particle states j with energies E_j exist (normally there is degeneracy, which means that various states have the same energy), and what we want to know is which rules the particles follow in occupying those states.

We have already seen that there is a great difference in the behaviour of bosons and fermions. While many bosons can occupy one and the same state, a fermion claims a state for itself alone. So the statistical rules will distinctly differ for both types of particle. Furthermore, there will also be a difference between the behaviour of massive bosons and photons, since the total number of particles has a given finite value in the first case, whereas it is unlimited, in principle, in the second case.

A characteristic quantity that reflects the statistical properties of the particles is the average occupation number for any state j, $\bar{n}_j$, as a function of the corresponding energy E_j. Before calculating this function, we should clarify the state concept. Let us begin with photons. For a radiation field, the state j has to be identified with a selected field mode, i.e., it will be characterized by a wave vector $\boldsymbol{k}$ and a certain polarization. This mode can be occupied by $n = 0, 1, 2, \ldots$ photons with energy $h\nu$, where the frequency ν is related to $|\boldsymbol{k}| = k$ through $\nu = ck/2\pi$. In the case of a gas of massive particles, the state is specified by a momentum vector and the spin direction, and the corresponding energy is the kinetic energy of the particle. In the quantum mechanical description states of sharp momentum $\boldsymbol{p}$ (momentum eigenstates) are plane waves, const $\exp(\mathrm{i}\boldsymbol{k}\boldsymbol{r})$, where $\boldsymbol{p} = \hbar\boldsymbol{k}$.

The average occupation number is readily calculated for photons. We can actually apply Boltzmann statistics, which is well known from classical thermodynamics. It relies on the general rule that the probability of finding a system in a given state i with energy E_i is

$$p_i = \mathrm{e}^{-\frac{E_i}{kT}}/Z \tag{8.7}$$

(k Boltzmann constant), where

$$Z = \sum_i \mathrm{e}^{-\frac{E_i}{kT}} \tag{8.8}$$

is the so-called partition function.

In the photon case the possible energy levels are $E_n = nh\nu$ so that (8.7) takes the form

$$p_n = \mathrm{e}^{-\frac{nh\nu}{kT}}\Big/\sum_m \mathrm{e}^{-\frac{mh\nu}{kT}}. \tag{8.9}$$

From this formula the average occupation number $\bar{n}$ for the mode under consideration is readily calculated to be given by

$$\bar{n} = \sum_n np_n = \frac{1}{\mathrm{e}^{\frac{h\nu}{kT}} - 1}. \tag{8.10}$$

On multiplication of this result by $h\nu$ we find the electromagnetic energy stored, on average, in a mode with frequency ν

$$\bar{E} = \frac{h\nu}{\mathrm{e}^{\frac{h\nu}{kT}} - 1}, \tag{8.11}$$

and this formula is in perfect agreement with Planck's radiation law, as will be shown below.

In deriving the results (8.10) and (8.11) we benefited from the fact that there exists no upper limit for the photon number, which allows us to extend the sums in (8.9) and (8.10) to infinity. However, when we are dealing with massive noninteracting particles with integer spin (a Bose gas), the total number of particles N will be finite, of course. This gives rise to a modification of (8.10) and (8.11) in the form

$$\bar{n}_j = \frac{1}{e^{\frac{E_j - \mu}{kT}} - 1}, \tag{8.12}$$

$$\bar{E}_j = \frac{E_j}{e^{\frac{E_j - \mu}{kT}} - 1}. \tag{8.13}$$

Here, $\bar{E}_j$ is the average energy associated with the state j, and μ, the so-called chemical potential, is a parameter whose value has to be determined from the requirement that the sum over the occupation numbers for all possible states equals the given total number of particles N

$$\sum_j \bar{n}_j = N. \tag{8.14}$$

It turns out that the chemical potential μ is always negative for a Bose gas (except for the photon case, where it vanishes).

For a Fermi gas one gets, of course, distinctly different results. Interestingly, it turns out that only the unity in the denominators in (8.12) and (8.13) changes its sign. So the average occupation number for the state j, which is now simply the probability to find a particle in this state, reads

$$\bar{n}_j = \frac{1}{e^{\frac{E_j - \mu}{kT}} + 1}, \tag{8.15}$$

and the average energy referring to the state j is given by

$$\bar{E}_j = \frac{E_j}{e^{\frac{E_j - \mu}{kT}} + 1}. \tag{8.16}$$

An immediate consequence of (8.15) is that the average occupation number is smaller than, or at best equal to, unity, as required by Pauli's exclusion principle. The chemical potential has again to be calculated from the requirement (8.14). For Fermi gases it takes on positive values at not too high temperatures, as we will see below.

In practice, we normally encounter the situation that a lot of states (modes) are simultaneously excited. (Only in resonators such as microwave cavities or laser resonators does it become possible to experimentally realize single-mode operation.) So what we need to know is their density. In fact, the latter is the same for photon as well as Bose and Fermi gases. This is because, quantum mechanically, all those particles are described by plane waves, as was mentioned above. The allowed propagation vectors k are fixed by appropriate boundary conditions. Apart from resonators whose metallic walls force the tangential component of the electric field strength to vanish, those conditions have a rather fictitious character: the plane waves are required to be periodic, with length L, in all three spatial directions. So the existence of a periodicity cube of volume $V = L^3$ is assumed. This periodicity condition selects an enumerably infinite number of modes whose k vectors satisfy the conditions

$$k_x = \frac{n_x}{2\pi L}, \quad k_y = \frac{n_y}{2\pi L}, \quad k_z = \frac{n_z}{2\pi L}, \tag{8.17}$$

where n_x, n_y, n_z are arbitrary integers.

We may visualize this situation by a cubic lattice with lattice parameter $2\pi/L$ in k space. Actually, we can associate any allowed k vector with one elementary cell of that lattice. For instance, we may assign to a k vector that cell whose left lower corner on the front side coincides with the vector's end point (its starting point being the origin in k space).

So, from a volume element in k space $dV_k = k^2 dk d\Omega$ ($d\Omega$ element of the solid angle) we obtain the mode density $\sigma_V(k)$ through division by the volume of a lattice cell $(2\pi/L)^3$

$$\sigma_V(k)dkd\Omega = g\frac{Vk^2 dkd\Omega}{(2\pi)^3} \quad (V = L^3). \tag{8.18}$$

We have introduced here an additional factor g that accounts for the degeneracy due to the spin properties of the particles. The degeneracy parameter is $g = 2$ for photons (thanks to the two polarization degrees of freedom), and it is $g = 2s + 1$ for massive particles, where s is the particle spin. So, specifically for electrons it equals 2 too.

What we are mainly interested in is the energy density. Then photons play an exceptional role since their individual energy depends linearly on k, $E = h\nu = \hbar ck$. Hence, starting from (8.18) we may define a spectral mode density $\sigma_V(\nu)$ through

$$\sigma_V(\nu)d\nu d\Omega = \frac{2V\nu^2 d\nu d\Omega}{c^3}. \tag{8.19}$$

On multiplication by (8.11) we find the average energy corresponding to the spectral range $\nu \ldots \nu + d\nu$ and the solid angle $d\Omega$ to be given by

$$\rho_V(\nu)d\nu d\Omega = \frac{2Vh\nu^3 d\nu d\Omega}{c^3} \frac{1}{e^{\frac{h\nu}{kT}} - 1}. \tag{8.20}$$

The appearance of the factor V in this equation is quite natural. In fact, the energy stored in any mode is distributed over the whole mode volume V. Hence it is a global rather than a local quantity. So, dividing (8.20) by V, $d\nu$ and $d\Omega$ leads us to the spatial and spectral energy density per unit solid angle of a radiation field at temperature T (in particular, of blackbody radiation)

$$\rho(\nu) = \frac{2h\nu^3}{c^3} \frac{1}{e^{\frac{h\nu}{kT}} - 1}, \tag{8.21}$$

and we are happy to see that we are freed from the fictitious periodicity volume. Actually, (8.21) is nothing but Planck's radiation law.

Considering now gases consisting of free massive particles (bosons or fermions), we must take into account that the energy of a particle is its kinetic energy

$$E = \frac{(\hbar k)^2}{2m}. \tag{8.22}$$

Utilizing again (8.18) and integrating over the solid angle (which yields a factor of 4π) we thus find the mode density for massive particles (the number of modes per unit energy interval)

$$\sigma(E) = \frac{gVm\sqrt{m}}{\sqrt{2}\pi^2\hbar^3}\sqrt{E}. \tag{8.23}$$

Let us now focus on Fermi statistics, which is of great practical importance since it determines the electronic properties of metals. Indeed, electrons in metals are approximately free so that they realize an electron gas, to a good approximation. What we have to determine is the chemical potential μ. According to (8.14) it satisfies the relation

$$\int \sigma(E)dE = \frac{V\sqrt{2m}\sqrt{m}}{\pi^2\hbar^3} \int \frac{\sqrt{E}}{e^{\frac{E-\mu}{kT}} + 1}dE = N. \tag{8.24}$$

It follows from (8.24) that the chemical potential depends on the spatial electron density N/V, as one would expect.

Unfortunately, there exists no closed-form solution to (8.24) so that we have to resort to numerical solutions, in general. In the interesting case of low temperatures

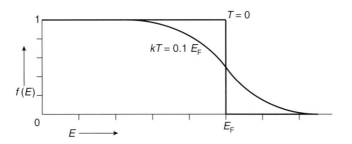

Fig. 8.1 Fermi–Dirac distribution for a three-dimensional gas. $f(E)$ = occupation probability as a function of the energy of the particle state.

T, however, an expansion with respect to T is successful which leads to the following approximate solution

$$\mu(T) = E_F - \frac{\pi^2}{12} \frac{(kT)^2}{E_F}, \qquad \frac{kT}{E_F} \ll 1, \tag{8.25}$$

where

$$E_F = \frac{\hbar^2}{2m} \left(\frac{3\pi^2 N}{V} \right)^{\frac{2}{3}} \tag{8.26}$$

is the so-called Fermi energy. Actually, the condition in (8.25) is fulfilled up to rather high temperatures (e.g., 1000 K for alkali metals).

Introducing the result (8.25) into (8.15), we get the average occupation number per state, as a function of the particle's energy, with T as a parameter (see Fig. 8.1). In the limit $T \to 0$, it becomes a step function that falls from unity to zero at $E = E_F$. This is just what one expects from Pauli's exclusion principle. Starting from the state with lowest energy, $E = 0$, the states with higher energies are filled up by one particle each, as long as particles are available. The highest energy reached in this way is just the Fermi energy. One observes from Fig. 8.1 that, with growing temperature, an increasing fraction of particles become excited to higher energies than E_F, which naturally gives rise to a deficiency of particles at energies lower than (and close to) E_F.

Hence it becomes obvious that the electron gas in metals differs distinctly in its properties from a classical gas (that is subjected to Boltzmann statistics) – one says it is degenerate – in a wide temperature range. In particular, the electronic contribution to the specific heat of metals is drastically reduced (by a factor of about 100 at room temperature) in comparison to what is expected for a classical gas. In fact, the specific heat of metals is readily measured (note that it is a *macroscopic* quantity), and the much too small measured results indicated a serious failure of classical theory.

8.4 Symmetry

One of the basic insights into the microscopic world provided by quantum theory is the general indistinguishability of particles. In fact, in classical theory all objects, irrespective of whether they are large or tiny, were considered as distinguishable, as a matter of course. How does quantum mechanics describe indistinguishability? An answer is readily found when we call to mind how quantum theory deals generally with uncertainties. It makes use of the (mathematical) possibility to *superpose* quantum states. So let us follow this recipe.

Let us consider, for simplicity, a two-particle system. What we might know is that one particle is in a state, say $|\xi\rangle$, and the other in a different state $|\eta\rangle$, however, we cannot tell, *in principle*, which particle is in which state. In other words, we cannot label the particles physically. Nevertheless, for mathematical purposes such a labelling will be necessary. It is only in this sense that we speak of particle 1 or 2 in the following. The two cases, (*i*) 'particle 1 in state $|\xi\rangle$ and particle 2 in state $|\eta\rangle$' and (*ii*) 'particle 2 in state $|\xi\rangle$ and particle 1 in state $|\eta\rangle$' cannot be distinguished physically. So we will write the total wavefunction as a superposition

$$|\psi\rangle = \frac{1}{\sqrt{2}}(|\xi\rangle_1 \, |\eta\rangle_2 + e^{i\alpha} \, |\xi\rangle_2 \, |\eta\rangle_1), \tag{8.27}$$

where the phase factor $\exp(i\alpha)$ has yet to be fixed. This is easily done by requiring that the *physical* state of the system remains the same when we interchange the labels 1 and 2. This means that the wavefunction (8.27) must be reproduced, up to a phase factor. This condition is fulfilled only for $\exp(i\alpha) = 1$ and $\exp(i\alpha) = -1$. So we arrive at two wavefunctions that are symmetrical and anti-symmetrical, respectively, with respect to an interchange of the labels 1 and 2:

$$|\psi\rangle_+ = \frac{1}{\sqrt{2}}(|\xi\rangle_1 \, |\eta\rangle_2 + |\xi\rangle_2 \, |\eta\rangle_1), \tag{8.28}$$

$$|\psi\rangle_- = \frac{1}{\sqrt{2}}(|\xi\rangle_1 \, |\eta\rangle_2 - |\xi\rangle_2 \, |\eta\rangle_1). \tag{8.29}$$

We still have to clarify which kind of particles require symmetrization or anti-symmetrization of the wavefunction. We find a strong indication when we specialize to the case $\xi = \eta$. Then the anti-symmetrical wavefunction (8.29) obviously vanishes, i.e., such a state does not exist, and this is nothing but what the exclusion principle says. So we will assign the anti-symmetrical wavefunction to fermions, and, consequently, the symmetrical wavefunction to bosons.

This rule also applies to the general case of systems formed by more than two (N) particles. What must be clearly stated, however, is what the symmetrical and anti-symmetrical wavefunctions look alike. The recipe to construct them is as follows.

Using the abbreviation

$$|\psi(1, 2, \ldots, N)\rangle = |\xi\rangle_1 \, |\eta\rangle_2 \cdots |\chi\rangle_N, \tag{8.30}$$

we have to write the symmetrical wavefunction as

$$|\psi\rangle_+ = A \sum_P P \, |\psi(1, 2, \ldots, N)\rangle, \tag{8.31}$$

where A is a normalization factor and the sum is over all possible permutations P of the set of labels $1, 2, \ldots, N$.

The anti-symmetrized wavefunction is required to change sign whenever two labels are interchanged. This is so when its form is chosen as

$$|\psi\rangle_- = B \sum_P \delta_P P \, |\psi(1, 2, \ldots, N)\rangle, \tag{8.32}$$

where δ_P equals 1 (-1) if the permutation P is equivalent to an even (odd) number of successive interchanges of label pairs. Actually, the mathematical structure in (8.32) is known from determinants. So anti-symmetrical wavefunctions can elegantly be written as determinants. The above symmetry postulates lead, in fact, to observable differences in interactions. This can be seen, in particular, when the scattering of identical particles is analyzed.

So let us go back to the previous discussion of scattering in Section 6.3. There we tacitly assumed that the target particles and the scattered particles are of different kinds. We will now investigate how the former analysis must be modified in the case of identical particles.

From (6.28) it becomes obvious that the exchange of the labels 1 and 2 amounts to replacing the relative coordinate x by $-x$. In Section 6.3 we assumed that (in the centre-of-mass system) the scattered particle (labelled 1) propagates in the ϑ, φ direction, whereas the target particle (labelled 2) moves in the opposite direction described by $\vartheta' = \pi - \vartheta, \varphi' = \varphi + \pi$. Hence the symmetrization of the wavefunction (6.38) gives us

$$|\psi\rangle_\pm = e^{ik_{in}x} \pm e^{-ik_{in}x} + [A(\vartheta, \varphi) \pm A(\pi - \vartheta, \varphi + \pi)] \frac{e^{ik|x|}}{|x|}. \tag{8.33}$$

So the differential scattering cross-section (6.40) has to be replaced with

$$\sigma = |A(\vartheta, \varphi) \pm A(\pi - \vartheta, \varphi + \pi)|^2. \tag{8.34}$$

Now, it should be noticed that we ignored spin properties in our previous scattering analysis. So we are certainly justified in applying (8.34), with the plus sign, to the scattering of identical spinless particles (i.e., particles with spin zero,

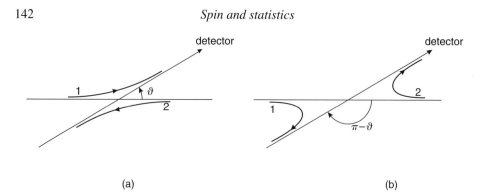

(a)   (b)

Fig. 8.2 Scattering of two particles. (a) The detector registers particle 1.
(b) It registers particle 2. When the particles are identical, the two events are
indistinguishable.

which are bosons), for instance to Coulomb scattering of α particles. (In this case,
A depends only on ϑ, owing to the radial symmetry of the Coulomb potential.)
Hence we have to choose the plus sign in (8.34), and we can write this equation as

$$\sigma = |A(\vartheta)|^2 + |A(\pi - \vartheta)|^2 + 2\mathrm{Re}\{A(\vartheta)A(\pi - \vartheta)\}. \tag{8.35}$$

Here, we can identify the first two terms with what one would expect from a naive
argument: the detector will count particles 1 as well as particles 2 propagating
in a given direction ϑ, φ (see Fig. 8.2), and, assuming both types of events to be
independent, one will add the corresponding probabilities. Actually, in the first case
particle 1 is scattered in the ϑ direction and in the second case in the $\pi - \vartheta$ direction.
Since the scattering amplitude A refers to particle 1, this explains that $A(\vartheta)$ and
$A(\pi - \vartheta)$ come into play.

   Quantum mechanics, however, tells us that this argument is erroneous, as
becomes obvious from the third term in (8.35), which has the form of an interference
term. Originating from the formal symmetry postulate for identical particles, it
describes an intrinsically quantum mechanical phenomenon which was named the
exchange effect. In Coulomb scattering, it is most pronounced at $\vartheta = \pi/2$ (which
means a scattering angle of $\pi/4$ in the laboratory system), where it doubles the
naively expected scattering cross-section.

   Now, what about scattering of particles with spin? Let us focus on spin-$\frac{1}{2}$ particles
such as electrons and protons. Normally, the probability that the spin changes its
orientation in the scattering process is negligible. So both the total spin and the
total angular momentum are separately integrals of motion. As a result, the total
wavefunction factorizes into a spin state and a wavefunction in coordinate space.
The latter is the wavefunction I discussed in Section 6.3. Since we are now dealing

with fermions, the total wavefunction is required to be anti-symmetrical. This leads us to two possible situations:

(*i*) The spins are in a singlet state

$$|\varphi\rangle_s = \frac{1}{\sqrt{2}} \left(\left|+\frac{1}{2}\right\rangle_1 \left|-\frac{1}{2}\right\rangle_2 - \left|+\frac{1}{2}\right\rangle_2 \left|-\frac{1}{2}\right\rangle_1 \right), \tag{8.36}$$

where $\left|+\frac{1}{2}\right\rangle, \left|-\frac{1}{2}\right\rangle$ describe states with spin components $+\frac{1}{2}$ and $-\frac{1}{2}$ with reference to a chosen quantization axis, respectively. Obviously, the spins are antiparallel, and (8.36) is an eigenstate of the total spin with eigenvalue 0. Since the spin state (8.36) is already anti-symmetrical, the spatial wavefunction must be symmetrical, i.e., (8.34) has to be taken with the plus sign.

(*ii*) The spins are in a triplet state which has total spin 1. (Actually, this state is threefold degenerate, corresponding to the three possible orientations of the total spin, the individual spins being parallel.) Since the triplet state is symmetrical, the spatial wavefunction has to take care of anti-symmetry. This means that (8.34) now applies with the minus sign.

9

Macroscopic quantum effects

9.1 Quantum mechanics casts its shadows before it

It might be felt amazing that quantum effects show up in our macroscopic world. We know miraculous phenomena such as superconductivity, for instance, which classical physics is basically unable to explain. This makes it impossible to separate nicely the microcosm from the macroscopic world through a clear-cut borderline, or a more or less well defined transition region at least. It should be noticed that classical physics, notably statistical thermodynamics, in certain cases already runs into trouble in what should be its domain of validity. This became obvious from observations, mostly at low temperatures, that manifestly differed from basic classical predictions.

Those discrepancies were found, above all, in thermodynamics. In particular, the specific heat capacity of materials with decreasing temperature did not obey the rule that followed from classical (i.e., Boltzmann) statistics. This rule states that the energy can be calculated by assigning to any constituent the average energy $kT/2$ (k Boltzmann constant) per degree of motional freedom. (In the case of vibrations the potential energy gives rise to an extra contribution $kT/2$.) For instance, in an ideal gas formed from atoms, any atom has three translational degrees of freedom. Hence each atom contributes the energy $3kT/2$, and we simply get the total energy of the sample on multiplication of this figure by the number of atoms. When the constituents of a gas are molecules, additional rotational and vibrational degrees of freedom come into play, which further enhance the average energy of a molecule.

Experience, however, revealed that vibrational and rotational degrees of freedom gave no contributions to the total energy at normal and lower temperatures. It seemed as if those degrees of freedom got 'frozen'. But why? A similar effect was observed in solids, where the thermal motion has the form of lattice vibrations. As a matter of fact, at low temperatures the vibrations contribute less than expected from the above-mentioned classical rule, and at very low temperatures their contribution even vanishes. Further, the specific heat capacity of the electrons in metals already becomes drastically reduced at room temperature (cf. Section 8.3).

Similarly, it turned out that classical theory was incapable of predicting correctly the spectral energy distribution of blackbody radiation. In fact, application of the classical rule to the radiation field led to a catastrophe. One has to identify the degrees of freedom with the field modes. Since both the kinetic and the potential energy contribute $kT/2$ to the energy of a vibration, on average, we have to multiply the mode density, to be taken from (8.19), by kT. Integrating over the full solid angle Ω we thus arrive at Rayleigh's radiation law giving us the spatial and spectral energy density of the field as

$$\varrho(\nu) = \frac{8\pi}{c^2} \nu^2 kT. \tag{9.1}$$

While this law agrees well with observations at small values of ν/T, it obviously becomes absurd at high frequencies. Since $\varrho(\nu)$ diverges as ν^2 for $\nu \to \infty$, the total energy (the integral of ϱ over ν) hopelessly diverges.

Actually, all attempts to deduce a satisfactory radiation law in the framework of classical theory were doomed to fail. Even Wien's semi-empirical law which avoided the 'ultraviolet catastrophe', eventually turned out to be distinctly at variance with measurements at very long wavelengths (about $10\,\mu m$). This dilemma led Planck, in what he called an 'act of despair', to resort to a quantization hypothesis that allowed him to improve Wien's law properly. It was only later that the revolutionary character of Planck's work was recognized and the date of his report at the 'Berliner Kolloquium' of the Physical Society, 14 December 1900, was celebrated as the birth hour of quantum theory. It should not be forgotten that, actually, very accurate *macroscopic* measurements had enforced this new kind of physical thinking.

After this introduction, I will discuss some of the most important quantum effects that manifest themselves in striking macroscopic phenomena.

9.2 Superconductivity

9.2.1 Experiment

In 1911 H. Kamerlingh Onnes at Leiden made a spectacular discovery which, actually, came as a complete surprise. Three years previously he had succeeded as the first one to liquefy helium. The availability of liquid helium, in turn, gave him an opportunity to cool down samples drastically. It turned out as a stroke of luck that he decided to investigate electric conductivity in those extreme conditions. As a convenient material he chose mercury (which is solid at those temperatures). What he observed was an abrupt decrease of its electrical resistance at $T = 4.3$ K, followed by a further dramatic fall below $3 \cdot 10^{-6}$ Ω at $T = 3$ K. (For comparison, the resistance of his mercury sample at $0°$ C, in the liquid state, was 172.7 Ω.)

This almost incredible phenomenon was termed superconductivity. It is almost needless to say that Kamerlingh Onnes was awarded the Nobel prize for his outstanding achievement (in 1913). Characteristic of superconductivity is the existence of a critical temperature T_c at which some kind of phase transition from a state of normal electric conductivity to a superconducting state takes place. In the latter, the d.c. resistance is practically zero. Once induced, an electric current persists for years in a superconducting coil. Hence superconductivity opens up undreamed-of possibilities in electrical engineering. For example, the strongest electromagnets known to man can be made (which, in particular, have found application in the construction of modern ring-shaped particle accelerators).

A peculiar feature of superconductivity is that the superconducting state is destroyed through magnetic fields, provided they exceed a critical magnetic field strength H_c that depends on temperature. At the critical temperature T_c, H_c equals zero, and it rises with falling temperature.

In subsequent decades, superconductivity was observed in many metallic elements as well as alloys and intermetallic compounds. The transition temperatures for conventional superconductors vary over a domain that ranges from values below 1 K to around 20 K.

However, the disappearance of electric resistance was not the only surprise that superconductors had up their sleeves. Actually, in 1933 Meissner and Ochsenfeld observed a curious magnetic effect when they cooled a sample below the transition temperature in the presence of a constant, weak magnetic field. They found that the magnetic induction B, residing in the interior of the sample, completely disappeared. This happened abruptly when the temperature had fallen below the transition temperature. Strictly speaking, B penetrates the superconductor for a very short distance only. This so-called penetration length is of the order of 100 nm for most superconducting materials.

The Meissner effect can be explained by assuming that electric currents are circulating in the probe very close to its surface. They generate a homogeneous magnetization within the superconductor that compensates the internal B field produced by the applied magnetic field. To this end, the magnetization must obviously be opposite to B.

Accordingly, when observing, at a given temperature below the critical temperature T_c, the magnetization M of a superconducting sample in the form of a long cylinder, as a function of an applied longitudinal magnetic field H, one finds the following result. Starting from $H = 0$, M grows linearly with H. However, when H reaches a critical value H_c, M falls abruptly to zero, and at $H > H_c$ the sample becomes normally conducting. This behaviour is characteristic of a class of superconductors named type I. There exists, however, a second class referred to as type II, which exhibits a distinctly different magnetization curve: when the applied magnetic field has reached a first critical value H_{c1}, M begins to fall, however not abruptly, but smoothly until it reaches zero at a second critical value H_{c2} of the magnetic field. For $H > H_{c2}$ superconductivity again disappears.

9.2.2 Theory

For a long period, superconductivity opposed a physical explanation. The origin of electric resistance is normally collisions between the conduction electrons and lattice vibrations (in quantum theory these are phonons, the analogues of photons). In those collisions the electrons lose some of their kinetic energy through conversion into vibrational energy of the ions so that heat is produced. In addition, lattice defects, such as impurities, dislocations and disorder, give rise to electron scattering. For $T \to 0$, the phonons disappear. So one expects a residual resistance due to lattice defects to be present at very low temperatures. The observed abrupt complete disappearance of resistance, however, finds no explanation.

The first step towards a theory of conductivity was taken by the brothers F. and H. London, who in 1935 reproduced the Meissner effect theoretically. In 1950, this theory was successfully developed further by Ginzburg and Landau (1950). They introduced an order parameter $\psi(x)$ through the relation

$$|\psi(x)|^2 = n_s(x), \tag{9.2}$$

where n_s is the local concentration of the superconducting electrons. For the function ψ they derived an equation that resembles the (stationary) Schrödinger equation; however, it has an additional nonlinear term

$$\left[\frac{1}{2m'} \left(\frac{\hbar}{i} \mathbf{grad} - \frac{e'}{c} \mathbf{A} \right)^2 - \alpha + \beta |\psi|^2 \right] \psi = 0. \tag{9.3}$$

The superconducting electric current density is given by

$$\mathbf{j}_s = \frac{e'}{2m'} \psi^* \left(\frac{\hbar}{i} \mathbf{grad} - \frac{e'}{c} \mathbf{A} \right) \psi + \text{c.c.} \tag{9.4}$$

Here, $\mathbf{A}$ is the vector potential of an electromagnetic field being present, α and β are positive phenomenological constants, and m', e' are the mass and the electric charge of the superconducting charge carriers. According to the BCS theory explained below, those carriers are electron pairs. So we have $m' = 2m$, $e' = 2e$, where m and e are the electronic mass and charge.

The mathematical structure of the superconducting current density (9.4) is well known in quantum theory. With missing prefactor e' and the identification of $e' = e$ and $m' = m$, it is the probability current density j, in the presence of an electromagnetic field, that ensures the probability density for an electron, $\rho = |\psi|^2$, to be a locally conserved quantity, as expressed by the continuity equation

$$\frac{\partial}{\partial t}\rho(\boldsymbol{x}, t) + \mathrm{div}\boldsymbol{j}(\boldsymbol{x}, t) = 0. \tag{9.5}$$

In fact, the latter says that the temporal change of the probability of finding the electron in a volume element $\Delta\tau$ is connected with a net flow of probability into, or out of, $\Delta\tau$. By the way, a conservation law of this kind also exists in classical electrodynamics. It describes the local conservation of electromagnetic energy with j being the density of the energy flux (Poynting vector).

Equation (9.3) became known as the Ginzburg–Landau equation. It proved to be very successful in explaining the macroscopic properties of superconductors. In particular, it was shown by Abrikosov that it accounts for the existence of the two types of superconductor that differ in their magnetization curves (see above). However, the Ginzburg–Landau equation has phenomenological character. So theorists were eager to develop a microscopic theory that provided insight into the physical mechanism underlying superconductivity.

This goal was achieved in 1957 by Bardeen, Cooper and Schrieffer (1957). Their theory, known as BCS theory, is based on the concept of electron pairs, the so-called Cooper pairs. Such a pair is formed of two conduction electrons with precisely opposite wave vectors and opposite spins. The two electrons interact with one another through an exchange of phonons. The physical picture is that a first electron slightly distorts the ionic lattice. This disturbance develops and propagates in the lattice, thus affecting a second electron at a distant position. The described interaction is attractive and hence leads to binding. Actually, the members of a Cooper pair are widely separated in space (up to 10^4Å)!

The pair concept gives us the clue to a physical understanding of superconductivity. In the ground state of a superconductor all electrons are paired. Excitation of the electron gas needs to 'break up' Cooper pairs. To this end, a finite energy Δ (the energy gap predicted by BCS theory) must be fed into a pair. This has the consequence that an excitation actually does not take place as long as the average thermal energy of a lattice oscillation kT (k Boltzmann constant) is smaller than Δ. So the electron gas will be in its ground state not only at $T = 0$, but also at somewhat higher temperatures. Now, the mystery of superconductivity is unveiled through the recognition that Cooper pairs (which have total momentum zero) cannot exchange momentum with lattice vibrations. (This follows from energy and momentum conservation.) So the above-mentioned conventional damping mechanism for an electric current induced by an applied electric field is 'switched off'. The electric current flows 'for ever'!

9.2.3 Magnetic flux quantization

Let us now come back to the function $\psi(x)$ that was introduced by Ginzburg and Landau as an order parameter. A deeper understanding is provided by an analogy with electrodynamics. In the quantum picture, an electromagnetic field is, roughly speaking, a gas of photons with different wave vectors and polarizations. On the other hand, there exists a classical description of the electromagnetic field through a classical wavefunction $\psi_{el}(x)$ that has the meaning of the electric field strength. (For simplicity, we consider linearly polarized light.) Hence we might identify the order parameter $\psi(x)$ with a 'classical' wavefunction of a system of paired electrons. In fact, (9.2) has an equivalent in electrodynamics, where $|\psi_{el}(x)|^2$, when appropriately normalized, gives us the probability density for detecting a photon. However, a wavefunction also possesses a phase. Its physical relevance is well known in electrodynamics, where it determines the interference behaviour. So we will ask whether the phase also has physical importance in superconductivity. In the following I will show that this is indeed the case.

We learn from (9.4) that the superconducting current density is determined by the phase; strictly speaking, its spatial dependence. In fact, when we assume a constant density n_s of superconducting pairs we may write

$$\psi(x) = \sqrt{n_s}e^{i\theta(x)}, \tag{9.6}$$

and (9.4) takes the form

$$\boldsymbol{j}_s = \frac{e'}{m'}n_s(\hbar\,\mathbf{grad}\,\theta - \frac{e'}{c}\boldsymbol{A}). \tag{9.7}$$

Let us now investigate the $\boldsymbol{B}$ field that exists within a superconducting ring. To this end, we consider a closed path $\mathcal{C}$ within the ring that keeps sufficient distance from the surface. In Section 9.2.1 we learned, in connection with the Meissner effect, that in the interior of a superconducting sample no electric current is flowing. Hence it follows from (9.7) that the relation

$$\boldsymbol{A} = \frac{\hbar c}{e'}\mathbf{grad}\,\theta \tag{9.8}$$

must hold at every point lying on the path $\mathcal{C}$.

Of physical interest is the contour integral over A since it gives us the magnetic flux ϕ. This is readily seen when we notice that $\boldsymbol{B} = \mathbf{rot}A$, by definition, and make use of Stokes' theorem

$$c \oint A_s \mathrm{d}s = s \int (\mathrm{rot}A)_n \mathrm{d}\sigma = s \int B_n \mathrm{d}\sigma = \phi. \tag{9.9}$$

Here, A_s, $\mathrm{d}s$ are the tangential component of A and the length element with respect to the path, and the subscript n denotes the normal component with respect to the surface S, $\mathrm{d}\sigma$ being the surface element.

So what we have to do according to (9.9) and (9.8) is to calculate the contour integral over $\mathbf{grad}\ \theta$. The result is

$$c \oint (\mathrm{grad}\ \theta)_s \mathrm{d}s = \theta_2 - \theta_1, \tag{9.10}$$

where θ_1 is the value of θ at the starting point of the integration and θ_2 is the value θ takes after a round trip along $\mathcal{C}$. Adopting the view that ψ is something like a 'classical' wavefunction, we will require it – as in the Schrödinger theory – to be a single-valued function. This implies that θ_2 and θ_1 can differ by a multiple of 2π only. Hence the combination of (9.9), (9.8) and (9.10) gives us

$$\phi = \frac{2\pi\hbar c}{2e} m \qquad (m \text{ integer}), \tag{9.11}$$

where we have identified e' with the charge of a Cooper pair, $e' = 2e$.

This is really a surprising result. It says that the flux which is a *macroscopic* quantity, is quantized, the flux quantum being

$$\phi_0 = \frac{2\pi\hbar c}{2e} = 2.0678 \cdot 10^{-7}\ \text{Gauss cm}^2 = 2.0678 \cdot 10^{-15}\ \text{Tesla m}^2. \tag{9.12}$$

It should be noticed that the flux through the superconducting ring is the sum of the flux originating from external sources and the flux produced by the superconducting current circulating near the surface of the probe. So, the latter flux must adjust to the former such that quantization of the total flux is ensured, as required by (9.11).

So we have indeed revealed a macroscopic quantum effect par excellence. The idea that the world can be separated nicely into a microcosm governed by quantum laws and a macrocosm subjected to *solely* classical laws cannot be upheld!

9.2.4 Josephson effects

In Section 6.4 the tunnelling effect, a specific quantum phenomenon, was explained, and it was mentioned that it finds interesting applications in semiconductor physics.

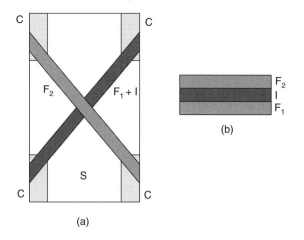

Fig. 9.1 Realisation of a Josephson junction. (a) Geometry. S = substrate; C = contact; $F_1 + I$ = superconducting film covered by an insulating layer; F_2 = superconducting film deposited on the top. (b) Layer structure at the crossing.

In 1962 B. D. Josephson, at the time still a student, came up with the lucky idea of investigating tunnelling of Cooper pairs through a thin barrier formed by an insulating layer (see Josephson (1974)). He made spectacular predictions that became famous as Josephson effects and were, in fact, experimentally verified.

First, Josephson predicted a superconducting current to flow through the insulating layer even when no voltage is applied, provided the layer is extremely thin (about $10\,\text{Å}$). A typical experimental set-up is shown in Fig. 9.1. Two superconducting metallic films and an insulating layer are vapour deposited, one upon the other, on a substrate like glass. The insulating layer became known as a Josephson junction.

Supposing that quantum tunnelling of Cooper pairs exists, we may analyze the resulting current by adopting a Schrödinger equation for two Cooper pairs separated by the layer. (I follow here the treatment by Kittel (1986).) Specifically, we describe an electron pair on one side of the junction layer by a quantum mechanical wavefunction ψ_1 and a pair on the other side by ψ_2. Postulating the existence of an interaction between those pairs that tends to induce a transfer through the junction we may choose, as a simple form of evolution equation, the following coupled Schrödinger equations:

$$i\hbar\frac{\partial \psi_1}{\partial t} = \hbar T \psi_2, \qquad i\hbar\frac{\partial \psi_2}{\partial t} = \hbar T \psi_1, \tag{9.13}$$

where $\hbar T$ describes the transfer interaction.

Splitting the wavefunction again into a (positive) amplitude and a phase factor,

$$\psi_1(t) = \sqrt{n_1(t)}e^{i\theta_1(t)}, \qquad \psi_2(t) = \sqrt{n_2(t)}e^{i\theta_2(t)}, \tag{9.14}$$

we get from (9.13), after a little algebra, the following relations

$$\frac{\partial n_1}{\partial t} = 2T\sqrt{n_1 n_2}\sin\delta = -\frac{\partial n_2}{\partial t}, \tag{9.15}$$

$$\frac{\partial\theta_1}{\partial t} = -T\sqrt{\frac{n_2}{n_1}}\cos\delta, \qquad \frac{\partial\theta_2}{\partial t} = -T\sqrt{\frac{n_1}{n_2}}\cos\delta, \tag{9.16}$$

where the abbreviation $\delta = \theta_2 - \theta_1$ has been introduced.

Now, when the two superconductors are identical, as we will assume, n_1 and n_2 will be approximately equal. Then (9.16) implies that $\partial\delta/\partial t = 0$, which means that δ is time-independent. Furthermore, (9.15) indicates that a pair vanishing at one side of the junction will appear at the other side. Now, the passage of a pair through the junction will contribute to a tunnel current J which, therefore, will be proportional to $\partial n_2/\partial t$ or, equivalently, $-\partial n_1/\partial t$, so that we obtain from (9.15)

$$J = J_0\sin\delta, \tag{9.17}$$

where J_0 is a constant proportional to T. This equation means that we observe (without a voltage being applied to the junction!) a d.c. current whose actual value lies between $-J_0$ and $+J_0$, depending on the phase difference δ across the junction layer. This is the d.c. current Josephson effect.

We have seen that no voltage is required to generate a superconducting current through the junction. So one will wonder what will happen when a voltage V is actually present. Let us discuss this problem in our simple model. The voltage across the junction gives rise to an energy step $2eV$ for an electron pair. We can take this into account by assigning the potential energy eV to one pair, and $-eV$ to the other. Thus the equations of motion (9.13) have to be replaced by

$$i\hbar\frac{\partial\psi_1}{\partial t} = \hbar T\psi_2 - eV\psi_1, \qquad i\hbar\frac{\partial\psi_2}{\partial t} = \hbar T\psi_1 + eV\psi_2. \tag{9.18}$$

Proceeding along the same lines as before, we arrive at the following results: (9.16) still applies, only the phase difference $\delta = \theta_2 - \theta_1$ now becomes time dependent

$$\delta(t) = \delta(0) - \frac{2eV}{\hbar}t. \tag{9.19}$$

So we observe an a.c. current

$$J(t) = J_0\sin[\delta(0) - \omega t], \tag{9.20}$$

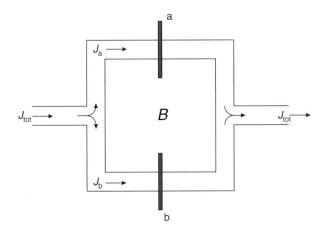

Fig. 9.2 Superconducting loop with two Josephson junctions a and b. J_a, J_b, J_{tot} = electric currents; $\boldsymbol{B}$ = magnetic field perpendicular to the drawing plane.

where

$$\omega = \frac{2eV}{\hbar}. \tag{9.21}$$

This is the a.c. current Josephson effect. For example, a voltage $V = 1\,\mu\text{V}$ produces an a.c. current with (circular) frequency $\omega = 483.6$ MHz. Equation (9.21) provides a novel opportunity to determine very precisely the value of $e/\hbar$ through measurement of a voltage and a frequency.

It is interesting to note that a special superconducting device was developed that proved to be a highly sensitive instrument for measuring magnetic fields. The measurement technique relies on *macroscopic* quantum interference.

The apparatus contains two Josephson junctions mounted in parallel, as depicted in Fig. 9.2. The two superconductors form a loop, and we suppose an orthogonal $\boldsymbol{B}$ field to be present. As we have seen above, the corresponding magnetic flux ϕ through the loop is given by the phase difference associated with a round trip in the superconducting loop

$$\theta_2 - \theta_1 = \frac{2e}{\hbar c}\phi. \tag{9.22}$$

(This relation follows directly from (9.11) by undoing the substitution $\theta_2 - \theta_1 = 2\pi m$.)

What is new compared with the former configuration is the presence of two Josephson junctions. According to (9.17) they give rise to d.c. Josephson currents J_a and J_b that flow through the junctions a and b, respectively.

Let us now evaluate the phase difference in (9.22). Since it refers to a round trip, we have to subtract the phase differences along the paths through the junctions a and b, δ_a and δ_b, respectively

$$\theta_2 - \theta_1 = \delta_b - \delta_a. \tag{9.23}$$

In the absence of a magnetic flux ϕ, δ_a and δ_b will be equal, $\delta_a = \delta_b = \delta_0$. This is not so, however, when a flux ϕ exists. Then (9.22) must be fulfilled; this is achieved by putting

$$\delta_b = \delta_0 + \frac{e}{\hbar c}\phi, \qquad\qquad \delta_a = \delta_0 - \frac{e}{\hbar c}\phi. \tag{9.24}$$

(The different sign in these equations is easily understood by observing that the currents J_a and J_b have the magnetic field on the right and the left, respectively.) Applying the formula (9.17) to both J_a and J_b, we find the total current to be given by

$$J_{\text{tot}} = J_a + J_b = J_0 \left[\sin\left(\delta_0 - \frac{e}{\hbar c}\phi \right) + \sin\left(\delta_0 + \frac{e}{\hbar c}\phi \right) \right] = 2J_0 \sin \delta_0 \cos \frac{e\phi}{\hbar c}. \tag{9.25}$$

Hence the total current is an oscillating function of ϕ which has maxima at

$$\phi = \frac{\pi \hbar c}{e} m \qquad (m \text{ integer}). \tag{9.26}$$

This relationship opens a novel opportunity to measure very weak magnetic fields, including those emanating from the human brain. The corresponding devices became known as SQUIDs (superconducting quantum interference devices). Actually, they are the most widely used superconductive devices.

9.3 Quantum Hall effect

9.3.1 Classical Hall effect

The classical Hall effect is a manifestation of the Lorentz force acting on a moving charge in the presence of a magnetic field. Let us consider the following typical situation. In an electric conductor with rectangular cross-section, a d.c. electric current is flowing in the x direction under the action of an electric field strength E_x. A homogeneous magnetic field $\boldsymbol{B}$ is applied in the z direction. Then, an electron moving with velocity v_x, as part of the electric current, is subjected to the Lorentz force

$$\boldsymbol{F} = -\frac{e}{c}\boldsymbol{v} \times \boldsymbol{B} \tag{9.27}$$

which, obviously, is orthogonal to both the velocity and the magnetic field. In our case, the Lorentz force points in the y direction,

$$F_y = \frac{e}{c} v_x B_z. \tag{9.28}$$

So, what will happen? After switching on the electric field, the electrons will drift sideways (in the y direction). As a result, electrons will accumulate at one lateral face of the conductor, while on the opposite side a deficit of electrons (and hence an excess of positively charged ions) emerges. Thus a transverse electric field E_y arises that counteracts the Lorentz force. Actually, in the evolving stationary state, this field will compensate the Lorentz force so that the electrons move straight forward in the x direction. The interesting point is the presence of the transverse field E_y, and this is the Hall effect.

9.3.2 Magnetic resistance in a two-dimensional channel

The Hall effect exhibits amazing quantum features when we study it in a two-dimensional system, that is to say in a very thin layer. In a typical set-up, the latter is the contact surface between a semiconductor (Si) and an insulating oxide layer (silicon dioxide SiO_2). On its upper side the oxide layer has a metallic electrode (see Fig. 9.3). This configuration, metal–oxide–semiconductor (MOS), has found application in transistor techniques. A voltage, the so-called gate voltage, is applied between the metal and the semiconductor. It gives rise to an enrichment of charge carriers at the oxide interface, thus producing a conductive channel.

Let the interface be the x, y plane. We assume a magnetic field $\boldsymbol{B}$ to be applied orthogonal to the interface (in the z direction), in addition to an electric field in the x direction, E_x. As was explained in the preceding section, the two fields cause a drift of the electrons (in the y direction). So when – unlike the device considered in Section 9.3.1 – we mount appropriate electric contacts, we will observe an electric current flowing in the y direction, driven by the electric field in the x direction.

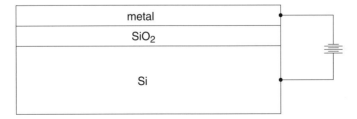

Fig. 9.3 Metal–oxide–semiconductor (MOS) layer structure with applied gate voltage.

A theoretical analysis shows (cf. Kittel (1986)) that this current, in the case of strong magnetic fields and low temperatures, is given by

$$I_y = \frac{V_x}{\rho_H},\tag{9.29}$$

where V_x is the voltage applied in the x direction and ρ_H is the Hall resistance

$$\rho_H = \frac{B}{vec}.\tag{9.30}$$

Here, v is the electron surface density (number of electrons per unit area) produced by the gate voltage.

But how does quantum theory come into play? The answer is that an (approximately free) electron subjected to a magnetic field is allowed, by quantum theory, to move along selected trajectories only. This already follows from the Bohr–Sommerfeld quantization condition

$$\oint p_s ds = (n+\gamma)2\pi\hbar \qquad (n \text{ integer}).\tag{9.31}$$

The contour integral is along the trajectory, p_s is the tangential component of the electron's (total) momentum, and γ ($0 < \gamma < 1$) is a correction term.

Now, in the presence of a magnetic (more generally, an electromagnetic) field, the latter gives a contribution eA/c to the total momentum, where A is the vector potential. Taking this into account and making use of Stokes' theorem (see (9.9)), one readily derives from (9.31) that the magnetic flux through a surface bounded by an allowed trajectory,

$$\Phi = \int B_n d\sigma,\tag{9.32}$$

is quantized according to

$$\Phi = (n+\gamma)\frac{2\pi\hbar c}{e}.\tag{9.33}$$

The corresponding quantum states became known as Landau levels.

Starting from (9.33), it is not difficult to show (cf. Kittel (1986)) that the degeneracy of a Landau level is

$$D = \rho B, \qquad \rho = \frac{eL^2}{2\pi\hbar c}\tag{9.34}$$

(independent of *n*), where *L* is the length of the probe assumed to have a quadratic form.

Given a total number of electrons *N*, at low temperatures they will fill up the Landau levels, beginning with the lowest level *n* = 1. Normally, there will be a highest level *n* = *s* that is completely filled, while the following level, *n* = *s* + 1, is only partly filled. However, special cases are also possible in which the highest level *s* is completely filled and the following level *s* + 1 is totally empty. In such singular cases we have, according to (9.34),

$$\rho B s = N. \tag{9.35}$$

Introducing the electron density $\nu = N/L^2$, we obtain from (9.35) and (9.34)

$$\nu = \frac{se}{hc} B. \tag{9.36}$$

Inserting this result into the expression (9.30) for the Hall resistance, we see that the latter becomes quantized

$$\rho_H = \frac{h}{e^2} \frac{1}{s} \qquad (s = 1, 2, 3, \ldots). \tag{9.37}$$

Apart from the factor 1/*s*, the Hall resistance is actually an *absolute* constant!

As amazing as this prediction is, it seems to be of less experimental relevance. To utilize it would require one to identify experimentally the singular physical situations it applies to. This, however, does not seem possible. The only thing that can be measured is the Hall resistance as a function of the gate voltage (or, alternatively, the magnetic field strength), and such a curve is expected to be continuous.

Therefore, it made a sensation when von Klitzing, Dorda and Pepper (1980) found the dependence of the Hall resistance on the gate voltage to exhibit clearly discernible plateaux that *precisely* (to nearly one part in a billion) corresponded to the quantized values (9.37) of the Hall resistance. No wonder, then, that von Klitzing was awarded the Nobel prize for this achievement in 1985. His findings, which became known as the von Klitzing or integral quantum Hall effect, are, in fact, not understandable when the two-dimensional crystal is assumed to be ideal. It turned out, however, that the characteristic plateaux can be explained as a result of imperfections. The basic mechanism is that disorder and impurities give rise not only to a broadening of the Landau levels but also to localized electron states adjacent to the (broadened) Landau levels. Those localized states are separated from the delocalized states around the centres of the Landau levels by so-called mobility edges. Now, electrons in localized states do not contribute to electron transport processes. Hence, the Hall resistance remains constant, as indicated by the plateaux, when the additional electrons produced through increasing the gate voltage fill up localized states.

Experimentally, the quantum Hall effect is of great importance since it allows us to define an *absolute* resistance standard $\rho = h/e^2 \approx 25\ 812.8\ \Omega$. Moreover, since the value c for the velocity of light is *exactly* known by definition, an extremely precise independent determination of Sommerfeld's fine structure constant $\alpha = e^2/\hbar c$ becomes possible.

10

Quantum computing

10.1 Why quantum computing?

As we all know very well, conventional (electronic) computers have enormous power which, moreover, is constantly increasing. With their help gigantic quantities of data can be handled and evaluated. This is of crucial importance, for instance, in modern high energy physics experiments. On the other hand, computers can be used to model very complex systems with the aim of predicting, for example, future developments in our environment, notably climatic changes. Thanks to their incredible speed computers are, of course, destined to treat mathematical problems for which specific algorithms exist very efficiently. To mention a simple example, the digits of π are readily evaluated up to a very high order. Last, but not least, their potential to solve differential equations numerically, even when they are nonlinear, makes computers indispensable in modern research.

So, one might ask, is there any reason to look for fundamentally new ways of computing as they are seemingly offered by quantum processes? What advantages can be expected when the laws of quantum mechanics, rather than those of classical physics, are exploited for computation? Two answers were given: first, the simulation of complex quantum systems on classical computers meets serious difficulties that might be overcome with the help of quantum computers. This idea was suggested in 1982 by R. Feynman, and in the 1990s it was indeed shown by several groups of researchers that quantum mechanical systems for which no efficient simulation on a classical computer is known can be efficiently simulated on a quantum computer.

Second, and more surprising, it was shown for a few special cases that a quantum computer will be superior to a classical computer in solving selected mathematical problems, notably the problem of finding the prime factors of an integer. Also here the quantum computer excels in efficiency. The efficiency criterion is the time (more precisely speaking, the number of computational steps) needed for the calculation.

When this time grows as a polynomial with the size of the input, we speak of an efficient algorithm. On the other hand, when this growth is superpolynomial (typically exponential), the algorithm is considered inefficient. In this sense, the classical algorithm for factorizing integers, for example, is inefficient. For very large integers the calculation is so extremely time consuming that the problem, in fact, becomes intractable. To give you an idea of the critical order of magnitude of the integer, let me mention that the best known method, the 'number field sieve', requires about 10^{18} computational steps to factorize a number of 130 decimal digits (see Steane (1998)). At an assumed speed of 10^{12} operations per second, the computation would take 42 days. However, when we double the number of decimal digits, the computation time would reach the fantastic value of a million years! On the other hand, Shor (1994) devised a quantum algorithm for the factorisation problem which would require about 2×10^{10} logic gates per run to factorize numbers with 130 decimal digits (see Steane (1998)). This would certainly be an extreme challenge to an experimentalist. The decisive point, however, is that a classical solution becomes impossible, if we double the number of digits to 260, whereas an ideal quantum computer would have to run only eight times longer than before.

So it is no wonder that the prospects of quantum computing fascinate both mathematicians and physicists. There are actually two different topics of research: (*i*) new efficient algorithms have to be devised that will run (only) on quantum computers, and (*ii*) physical concepts of quantum computers have to be discussed that, hopefully, might lead to a realization. Admittedly, in view of the desired complexity there is no great hope that a quantum computer will be constructed some day that can actually beat a classical computer. Nevertheless, it is an exciting task to study quantum mechanics from a completely novel point of view. Instead of striving to find out the physical laws governing microscopic systems and utilize this knowledge to develop technical devices, we now want to use the quantum laws to do calculations that are impossible to perform on conventional computers.

10.2 Qubits

Modern electronic computers make use of the digital representation of numbers. This means that any integer z is written as a sum of powers of two

$$z = a_0 2^0 + a_1 2^1 + a_2 2^2 + \dots . \tag{10.1}$$

The coefficients a_n ($n = 0, 1, 2, \dots$) can take the values 0 and 1 only, and such an entity is called a bit. This can be realized, for instance, by two states of opposite magnetization.

So the first task in building a quantum computer is to define a quantum mechanical counterpart of a bit, a so-called quantum bit, abbreviated qubit. It appears natural to think of quantum systems that can be in two orthogonal states. Examples are electrons whose spins may be oriented either in or opposite to a given direction (usually defined by a magnetic field), or photons that can be linearly polarized in one or the other of two orthogonal directions, or atoms (ions) with two separated long-lived levels. Similarly to classical computing, one will associate the basis states with the bit values 0 and 1, respectively. However, there is a fundamental difference between classical and quantum systems. It comes from the superposition principle which tells us that a qubit can be not only in one of the basis states which we will denote by $|0\rangle$ and $|1\rangle$, but also in any superposition state $\alpha\,|0\rangle + \beta\,|1\rangle$, where α and β are complex numbers that fulfil the normalization condition $|\alpha|^2+|\beta|^2 = 1$. For example, we have seen in Section 3.3.3 that arbitrary superposition states of two-level atoms can be excited by laser pulses of properly chosen intensity and duration. So, contrary to a classical bit, a qubit has a large potential of storing additional information. This is one of the features that make quantum computers attractive.

When making actual calculations, we certainly have to deal with many qubits. They define basis states in a high-dimensional Hilbert space, and it is those states that will be superposed in the general case.

Let us now consider a system of n qubits. We will introduce a convenient notation. The basis states are characterized by specifying whether the individual qubits are in the state $|0\rangle$ or $|1\rangle$. So we will write the basis states as $|j_1, j_2, \ldots, j_n\rangle$, where $j_k = 0, 1$ $(k = 1, 2, \ldots, n)$. For instance, the notation $|0, 1, 1, 0\rangle$ means that the first and the fourth qubit are in the state $|0\rangle$, and the second and the third qubit in the state $|1\rangle$. With respect to future calculations we will identify the numbers $j_1, j_2, \ldots, j_n$ with the dual representation of a number x,

$$x = j_1 2^{n-1} + j_2 2^{n-2} + \cdots + j_n 2^0. \tag{10.2}$$

So we can characterize the basis states by the numbers $x = 0, 1, 2, \ldots, 2^n - 1$, writing

$$|x\rangle = |j_1, j_2, \ldots, j_n\rangle. \tag{10.3}$$

An important point is that the linearity of quantum theory allows us to form superposition states $\sum_x a_x\,|x\rangle$, where a_x are arbitrary complex numbers subjected only to the normalization condition. So we can encode a host of information in our quantum system, in addition to the integers x. As will shortly be explained, this is indeed a quantum feature that can make a quantum computer much more efficient than a classical one.

However, when dealing with superposition states, we are facing a fundamental dilemma. Since a quantum computer will be conceived as a *single* system, the nonobjectifiability of quantum mechanics will give us trouble. While we can excite an individual quantum system in any superposition state, at least in principle, it is impossible to read the coefficients out. In fact, if we could do so, we could determine the quantum state of the system. As the EPR discussion shows, this would enable us to beat the causality principle – a horrible idea. So we must live with the fact that a quantum state of an individual complex system contains a host of information, most of which, however, remains concealed to us. Measurements can reveal only very restricted information. The reasons are twofold: first we can only choose a limited set of variables (corresponding to commuting Hermitian operators) to be measured at all, and second the outcomes of those measurements are normally not free from randomness. In fact, as we have learned in Chapter 3, a measurement gives rise to a reduction of the wavefunction, which means that just one out of a whole set of potential values becomes actual. This happens with a definite probability that can be calculated from the wavefunction. However, this is normally of little help when the measurement is carried out on a single system. (Exceptions are only probabilities very close to one or zero.)

Hence an algorithm to be implemented on a quantum computer should not only be more efficient than a competing classical algorithm, it is, moreover, required to lead to particular final states from which relevant information can actually be read out. In Section 10.4 it will be shown that this condition is indeed satisfied in two algorithms that became famous as demonstrations of the power of quantum computing. First, let me explain the working principles of a quantum computer.

10.3 How might a quantum computer work?

A computer, whether classical or quantum mechanical, is a physical system, and computing is achieved through physical processes. The fundamental difference between a classical and a quantum computer is that the former, as a classical system, obeys the laws of classical physics, whereas the latter makes extensive use of quantum laws. Quite generally, computation is a physical evolution process: the system starts from a certain initial state and ends up in a final state that represents the result of the computation. In the case of quantum computers, the evolution is known to be a unitary transformation (see Section 2.1). Hence, to make a computation on a quantum computer means to realize a specific unitary transformation. This is certainly no trivial task. Fortunately, we can benefit from the fact that any unitary transformation can be built up from rotations that bring single qubits into superposition states (see (10.6)), and elementary unitary transformations, so-called quantum gates (the analogues of gates in classical computers) that act on two qubits. Hence a quantum algorithm is specified by a sequence of rotations and gates, and the experimentalist's task is to realize this sequence of operations.

Among the quantum gates the so-called controlled-NOT gate (or C-NOT gate) is of special importance, since it was shown that any unitary transformation can be decomposed into rotations on a single qubit and controlled-NOT gates acting on two qubits, a control and a target qubit. A controlled-NOT gate has the following properties: it leaves the target qubit t unchanged, when the control qubit c is set to 0

$$|0\rangle_c |0\rangle_t \rightarrow |0\rangle_c |0\rangle_t, \qquad |0\rangle_c |1\rangle_t \rightarrow |0\rangle_c |1\rangle_t, \qquad (10.4)$$

and it causes a flip of the target qubit, when the control qubit is set to 1

$$|1\rangle_c |0\rangle_t \rightarrow |1\rangle_c |1\rangle_t, \qquad |1\rangle_c |1\rangle_t \rightarrow |1_c\rangle |0\rangle_t. \qquad (10.5)$$

To realize a controlled-NOT gate, one needs a physical mechanism that couples any target qubit to any control qubit. Cirac and Zoller (1995) found a suitable mechanism to be given by Coulomb repulsion, in their proposal of a quantum computer that is made up of cold ions stored in a linear Paul trap. This concept proved to be attractive for experimentalists. In fact, many of the techniques for preparing and manipulating quantum states of trapped ions have already been developed in the field of precision spectroscopy, and we mentioned that entanglement between two trapped ions was produced (see Section 4.3.2) and quantum teleportation was realized in a three-ion system (see Section 3.6.4). Hence the trapped ion scheme is indeed prospective. So, to get a realistic idea of how a quantum computer might work, as a physical system, let us have a closer look at the Cirac–Zoller proposal.

Dealing with ions, the two states of a qubit, $|0\rangle$ and $|1\rangle$, are realized by two long-lived energy levels. Working with alkaline earth ions, one may choose the state $|0\rangle$ as a sublevel of the $S_{\frac{1}{2}}$ ground state, and the state $|1\rangle$ as a sublevel of the excited $D_{\frac{5}{2}}$ state. Those two states are coupled through a quadrupole transition. So the excited state is metastable. Nevertheless, with the help of lasers optical transitions between the two states can be induced. A second important point is that by utilizing the quantum jump technique explained in Section 3.3.2, level occupations can be reliably read out. With a properly tuned strong laser, resonance fluorescence can be excited exclusively between the ground state $|0\rangle$ and a suitable higher level, whereas no such possibility exists for the excited state $|1\rangle$. So the occurrence of a fluorescence signal indicates that the qubit is in the state $|0\rangle$, whereas the absence of such a signal tells us that the state $|1\rangle$ is occupied. In fact, this is the kind of measurement that enables us to read out the result of a computation.

A prerequisite of computation is, of course, the ability to address the ions individually (with lasers). This is achieved by storing them in a linear Paul trap. With the help of the optical cooling technique mentioned in Section 2.6.3, the ions can be arranged in a 'string-of-pearls' configuration. Then each is localized in a region whose dimensions are much less than optical wavelengths. On the other hand, the spacing between adjacent ions can be made conveniently large.

While the ions are tightly bound in the directions transverse to the trap axis, the trap potential along the axis will give them the freedom to oscillate about their equilibrium positions. Their coupling via Coulomb repulsion will, however, give rise to collective oscillations that are described by normal modes of the linear chain. In the simplest case, all the ions move in the same way (as if they were rigidly coupled). The eigenfrequency of this so-called centre-of-mass (CM) mode, ν_{CM}, is the smallest one. It equals that of a single ion. Typically, it is of the order of 100 kHz. The eigenfrequencies of the various modes are well separated. In particular, the eigenfrequency next to ν_{CM} is greater by a factor of $\sqrt{3}$, and the eigenfrequencies of the other modes are still larger. So one can discriminate between the modes, and it becomes possible to excite, or de-excite, just one of them, preferably the CM mode.

All possible manipulations of the ions for computational purposes make use of lasers acting on individual ions. Depending on the tuning of the laser frequency, the following two basic processes can be realized.

(*i*) Rotation of a qubit. When the laser is at resonance with the transition $|0\rangle \leftrightarrow |1\rangle$, it will not affect the ion's motion. So we face the situation we discussed already in Section 3.3.3. There it was shown that a resonant laser pulse with properly chosen intensity and duration produces any desired superposition state. According to (3.1)–(3.3), the action of the laser can be described by the following unitary operation, which is referred to as qubit rotation $\hat{V}(\theta, \varphi)$;

$$|0\rangle \rightarrow \cos(\theta/2) |0\rangle - ie^{i\varphi} \sin(\theta/2) |1\rangle \,,$$

$$|1\rangle \rightarrow -ie^{-i\varphi} \sin(\theta/2) |0\rangle + \cos(\theta/2) |1\rangle \,. \tag{10.6}$$

Here we have introduced the pulse area $\theta = \Omega t$ (assuming the pulses to be rectangular) and disregarded the free evolution of the basis states, i.e., we have used the so-called interaction representation. Of special importance will be $\pi/2, \pi$ and 2π pulses that correspond to $\theta = \pi/2$, etc.

Not only must the frequency of the laser be adjusted very precisely; its polarization must also be properly set. This is because the basis states are actually sublevels with given magnetic quantum numbers so that the laser transition is subjected to selection rules that ensure angular momentum conservation. To drive only the $|0\rangle \leftrightarrow |1\rangle$ transition, the laser beam must be chosen to be right-handed circularly polarized, for instance. Switching to left-handed circular polarization gives us an opportunity to drive a transition from $|0\rangle$ to a second excited level $|2\rangle$, which we will refer to as an auxiliary level. But what is the use of it? Certainly, we are not interested in actually populating level $|2\rangle$, since this would invalidate the qubit concept. However, by applying a left-handed polarized 2π pulse to the state $|0\rangle$ we will bring back the ion into this state, but nevertheless produce a subtle effect: as becomes obvious from (10.6), with $|1\rangle$ replaced by $|2\rangle$, the state $|0\rangle$ will change its sign. Importantly, the state $|1\rangle$ remains unaffected. Such an operation is indeed required. For instance, a simple NOT operation that interchanges $|0\rangle$ and $|1\rangle$ cannot be realized by a rotation: a π pulse with phase $\varphi = \pi/2$ causes the transition $|0\rangle \rightarrow |1\rangle$, $|1\rangle \rightarrow -|0\rangle$. So the false sign of $|0\rangle$ still has to be inverted.

(*ii*) Interaction with vibrations. When a laser is properly detuned, it will act not only on the internal states of an ion, but also on its vibrational motion. The underlying mechanism is that a moving ion 'feels' the (given) electric field strength residing at the actual positions of the centre of mass. This makes the interaction Hamiltonian depend on the elongation of the ion, which implies a coupling between the electric field and the ion's motion, in addition to the coupling to the internal degrees of freedom. When we are dealing with a linear chain of ions, the electric field will interact with the collective vibrations (normal modes), though the field is acting on just one ion. Let us specify the physical conditions in which this interaction can actually be observed. First, it has to be noticed that in the quantum mechanical description the vibrational modes have quantized energies. In perfect analogy to the electromagnetic field modes, the energy is a multiple of $h\nu_{\text{vib}}$ (h Planck's constant), where ν_{vib} is the eigenfrequency of the respective vibrational mode. The role of photons is now played by phonons, and we speak of the presence of n phonons ($n = 0, 1, 2, \ldots$) when the energy is just $nh\nu_{\text{vib}}$. Of vital importance for the Cirac–Zoller proposal are processes in which an ion becomes excited and one phonon is simultaneously annihilated. Energy conservation then requires that one laser photon is annihilated jointly with one phonon, their energies being converted altogether into the excitation energy of the ion,

$$h\nu_{\text{L}} + h\nu_{\text{vib}} = h(\nu_1 - \nu_0), \tag{10.7}$$

where ν_{L} is the laser frequency, and $h\nu_1$, $h\nu_0$ are the energies of the excited state $|1\rangle$ and the ground state $|0\rangle$. Equation (10.7) gives us the frequency condition

$$\nu_{\text{L}} = \nu_1 - \nu_0 - \nu_{\text{vib}}, \tag{10.8}$$

which indicates that the laser frequency must be red-shifted. As was mentioned above, the eigenfrequencies for the vibrational modes differ distinctly. This makes it possible to act on just one selected mode, preferably the CM mode, by properly setting the laser frequency.

But the inverse process must also be possible (which is, in fact, a general rule): a laser-induced transition from the higher level to the lower level that is accompanied by the creation of a phonon. (Formally, this follows from the Hermiticity of the interaction Hamiltonian.) So we are able to act, with properly tuned laser pulses, simultaneously on both the internal and the vibrational degrees of freedom of the ionic system. When directed on the μth ion, a laser pulse with pulse area θ and phase φ gives rise to the following transformation that is referred to as $\hat{U}_\mu(\theta, \varphi)$ operation

$$|0\rangle_\mu |1\rangle_{\text{vib}} \rightarrow \cos(\theta/2) |0\rangle_\mu |1\rangle_{\text{vib}} - \text{i} e^{\text{i}\varphi} \sin(\theta/2) |1\rangle_\mu |0\rangle_{\text{vib}},$$

$$|1\rangle_\mu |0\rangle_{\text{vib}} \rightarrow -\text{i} e^{-\text{i}\varphi} \sin(\theta/2) |0\rangle_\mu |1\rangle_{\text{vib}} + \cos(\theta/2) |1\rangle_\mu |0\rangle_{\text{vib}}.$$

(10.9)

Here, $|0\rangle_{\text{vib}}$ and $|1\rangle_{\text{vib}}$ are vibrational states with 0 or 1 phonons, respectively. From (10.9) it is readily seen that a π pulse ($\theta = \pi$) allows us to realize the special operations we considered before. It should be noticed that the expression for the Rabi frequency, and hence also the pulse area, is different for the operations $\hat{V}(\theta, \varphi)$ and $\hat{U}_\mu(\theta, \varphi)$.

As in the case of qubit rotations, the existence of an auxiliary level $|2\rangle_\mu$ can be utilized to change the sign of the state $|0\rangle_\mu$, without affecting the state $|1\rangle_\mu$. This is achieved with the help of a properly polarized 2π pulse. It should be observed, however, that this process takes place only when the vibrational mode is excited. Otherwise the (red-shifted) laser pulse will not interact with the ion. So we are able to change the sign of $|0\rangle_\mu$ *on condition* that one phonon is present. (From (10.9) one learns that the phonon is actually preserved.) We will see below that this sign inversion can be used as an important ingredient of a controlled-NOT gate.

The $\hat{U}_\mu(\theta, \varphi)$ operation (10.9) allows us to create a phonon in a selected vibrational mode, by applying a π pulse on any excited ion μ. Since all ions are involved in this vibrational motion, the phonon will modify, in turn, the action of a suitably chosen pulse sequence directed on any ion κ. So the vibrational mode plays the role of a data bus: the information that the μth ion is initially excited is encoded into a phonon, and it can be read out on any ion κ. This kind of information transfer is, in fact, crucial to quantum computing, opening the way to implement two-qubit gates. I will illustrate this by the example of the controlled-NOT gate.

The first step in implementing a controlled-NOT gate will be to a apply a red-shifted π pulse, $\hat{U}_c(\pi, -\pi/2)$, to the control qubit c. If it is in the excited state $|1\rangle_c$, the pulse will 'swap' the excitation for a phonon. What we now need is a sequence of pulses acting on the target qubit t that has the following effect: it inverts the population of the target qubit if a phonon is present. However, it leaves the target qubit unchanged in the absence of a phonon. Cirac and Zoller (1995) proposed to build up this sequence from three pulses: (a) a resonant $\pi/2$ pulse with phase $-\pi/2$, $\hat{V}_t(\pi/2, -\pi/2)$, (b) a red-shifted 2π pulse, $\hat{U}_t^{(\text{aux})}(2\pi)$, with modified polarization so that it interacts with an auxiliary level $|2\rangle_t$ (see above) and (c) a resonant $\pi/2$ pulse with phase $\pi/2$, $\hat{V}_t(\pi/2, \pi/2)$. From (10.6) it is readily seen that the two pulses (a) and (c), applied one after the other, reproduce the state of the target qubit. However, in combination with the pulse (b) (which changes exclusively the sign of the state $|0\rangle_t$) they produce the desired inversion of the target qubit. Note that the phonon is still present. So the control qubit can finally be set back, with the help of a red-shifted π pulse, to the initial state $|1\rangle_c$, whereby the phonon is annihilated. Hence the requirement (10.5) is fulfilled.

On the other hand, when the control qubit is initially set to $|0\rangle_c$, the first pulse $\hat{U}_c(\pi, -\pi/2)$ does not affect the control qubit. Since no phonon is produced, the pulse (c) has also no effect. So only the resonant pulses (a) and (c) actually interact with the target qubit, their combined action, however, leaving the state of the target qubit unchanged, as was mentioned above. So also the requirement (10.4) is met. Actually, experimentalists at Innsbruck succeeded in realizing a controlled-NOT gate along the lines just described (Schmidt-Kaler *et al.*, 2003).

Finally, let me say a few words on how the laser pulses are generated and manipulated. One starts with a cw laser that is coupled into a single-mode optical fibre. Making the laser light pass through a Pockels cell placed between crossed polarizers, the pulse duration – and hence also the pulse area – can be adjusted. A Pockels cell is also used for setting the polarization. Addressing a particular ion is achieved with the help of a focusing lens system. The possibility of switching between different ions is provided by electro-optical beam deflectors. The problem that is really open is the scalability of the experimental scheme. Will it be possible to manipulate many hundreds of ions such that the incredibly large number of computational steps prescribed by the algorithm are realized?

10.4 Effective quantum algorithms

In Section 10.2 it was pointed out that the quantum mechanical measurement process imposes severe restrictions on possible quantum algorithms. Actually, in the case of an ion chain computer an energy measurement is performed on each individual ion. As a result, we find any ion either in its ground state or in its excited state. Making use of the dual representation (10.2), we identify the set of measured data with a certain integer x that characterizes one of the basis states. This meagre result must indeed be enough to give us the full solution to the mathematical problem we are investigating! This is a perfectly new challenge to designers of algorithms, and one might doubt whether it can be mastered. Fortunately, this pessimistic view is not substantiated as some impressive examples of quantum algorithms show. Among them, great interest has been attracted by an algorithm devised by Shor (1994) that allows one to find the period of a periodic function. Its importance arises from the fact that factorization of a given prime number can be reduced to this problem. Let me describe the basic features of Shor's algorithm.

The function whose period shall be determined is defined by $f(x) = a^x$ modulo N, where N is the number to be factorized and the integer $a < N$ is chosen randomly. To encode the function $f(x)$ into the quantum computer assumed to be realized by an ion chain, we use two registers, one for the argument x (which we will call the x register) and one for the f values (which we will refer to as the y register). Now, owing to the modulo operation in the definition of $f(x)$ the f values are smaller than N. So the nearest integer n greater than $\log_2 N$ is the number of ions we need to implement the y register. When we study the function $f(x)$ in an x interval that extends beyond $N-1$, we have to choose $m > n$ ions for the implementation of the x register. As was explained in Section 10.2, we are thus able to encode 2^m numbers into the x register, and 2^n into the y register.

We start from the state in which all ions are in their ground states (and no vibration is excited)

$$|\psi_0\rangle_{x,y} = |0\rangle_x |0\rangle_y.$$ (10.10)

Next we produce the following superposition of the x states

$$|\psi_1\rangle_{x,y} = \frac{1}{\sqrt{2^m}} \sum_{x=0}^{2^m-1} |x\rangle_x |0\rangle_y.$$ (10.11)

This is achieved by applying a resonant $\pi/2$ pulse $\hat{V}(\pi/2, \pi/2)$ to each ion in the x register. In fact, one learns from (10.6) that such a pulse induces the transition $|0\rangle_\mu \to 2^{-1/2}(|0\rangle_\mu + |1\rangle_\mu)$ on the μth ion, and the multiplication of all those terms yields the superposition state for the x register in (10.11).

In the following step we encode the f values into the y register. To this end, we devise a network of logic gates acting on both registers that transforms any state $|x\rangle_x |0\rangle_y$ into the state $|x\rangle_x |f(x)\rangle_y$. Applying this network on the state (10.11), we thus produce the state

$$|\psi_2\rangle_{x,y} = \frac{1}{\sqrt{2^m}} \sum_{x=0}^{2^m-1} |x\rangle_x |f(x)\rangle_y.$$ (10.12)

This is certainly an unconventional way of representing a function. The link between x and $f(x)$ is established through entanglement, which is one of the most striking quantum features (see Sections 4.3 and 5.2). Not only are a large (or even a huge) number of f values 'stored' in just one quantum state – it is just $2^m f$ values and they are encoded in a system of only $m + n$ ions – even more amazing, they were all calculated in one go. So massive parallel processing has been achieved, which proves to be one of the decisive mechanisms that makes quantum computing so effective. You should realize that for large values of m, say $m = 100$, the storage capacity in question is really fantastic: you can store 2^{100} data in fewer than 200 ions!

As was mentioned in Section 10.1, the factorization problem becomes intractable on classical computers (at the present state of the art) when the order of the number to be factorized, N, exceeds 10^{130}. So it is this domain in which quantum computing might gain practical relevance. Hence the quantum computer, in fact, will have to deal with a gigantic amount of data (of order N). In comparison, the number of ions required to build up the quantum computer is rather modest. (Since $10^{130} \simeq 2^{300}$, this number will be about 600 altogether at the limit of current classical methods.) However, it must be borne in mind that the mentioned abundance of data, the f values, are actually concealed from us. You can never read them out! Only one of them becomes 'visible' in a measurement. A measurement on the x register would give us just a single value x_0 that is selected at random. According to the quantum mechanical rules for the description of the measurement process, the wavefunction (10.12) collapses to the state $|x_0\rangle_x |f(x_0)\rangle_y$, and a subsequent measurement on the y register yields finally the value $f(x_0)$. This is all the information we can extract through a direct measurement.

Fortunately, this is not the end of the story. When our aim is only to find the period of the function $f(x)$, we need not know all the f values, or at least a large part of them, as in a classical treatment. Actually we can benefit from the wealth of information we have no access to. The trick is to make a measurement on the y register! Then we will get, again by chance, one value of f, say $f(x) = u$. However, when $f(x)$ is a periodic function, more than one argument x will correspond to u. Hence the state (10.12) collapses to a superposition of the respective states $|x\rangle_x$ in the x register, multiplied by the state $|u\rangle_y$ in the y register. Owing to the factorization, there are no longer correlations between the two registers, and we can disregard the y register.

Now, when the period is given by r, the aforementioned superposition state has the form

$$|\psi_3\rangle_x = \frac{1}{\sqrt{M}} \sum_{j=0}^{M-1} |d_u + jr\rangle_x, \tag{10.13}$$

where d_u is a constant integer depending on the measured value u and it will be assumed that the number of terms in the sum, M, is large compared to unity.

Though this is a wonderful formula that makes the period transparent, we are still not happy. We are again not capable of reading it out! Fortunately, there is a way out of this dilemma. It is provided by an operation known as quantum Fourier transform. This transformation is defined as

$$|x\rangle \rightarrow \frac{1}{\sqrt{w}} \sum_{k=0}^{w-1} e^{i2\pi kx/w} |k\rangle \qquad (w = 2^m) \qquad (10.14)$$

for all values of x.

Applying the corresponding network of operations to the state (10.13) we get

$$|\psi_4\rangle_x = \frac{1}{\sqrt{wM}} \sum_{k=0}^{w-1} \sum_{j=0}^{M-1} e^{i2\pi(d_u+jr)k/w} |k\rangle. \qquad (10.15)$$

One recognizes that the sum over j yields a number with a large modulus (that is actually given by M) only when rk/w is an integer. In fact, in this case all terms of the sum have the same phase so that they 'interfere' constructively, whereas otherwise the 'interference' will be destructive. Hence, to a very good approximation, (10.15) can be replaced by

$$|\psi_4\rangle_x = \frac{1}{\sqrt{r}} \sum_{k} f(k) |k\rangle, \qquad (10.16)$$

where $|f(k)| = 1$ if k is a multiple of w/r, and $f(k) = 0$ otherwise. So measuring the final state of the x register, we will find one value k which must satisfy the relation $k = \lambda w/r$, where λ is an integer.

Now we can argue as follows. Under the assumption that λ and r have no common factor, the period r is readily determined: We cancel the ratio k/w down to an irreducible fraction which gives us λ and r separately. This result should be checked, and when the check fails (since the assumption is wrong), the whole algorithm must be repeated from the start. It has been shown that usually a number of repetitions much less than $\log r$ will suffice for success.

This is how Shor's algorithm works. We have seen that the fundamental quantum features such as the superposition principle (10.11), entanglement (10.12) and interference of probability amplitudes (10.15) were exploited. So the full power of quantum mechanics was summoned up to beat classical computers.

A quantum computer might also be employed to solve a quite different task. It is the problem of searching for a needle in a haystack, a proverbial example of a hopeless endeavour. A more timely example is to look for a particular telephone number in a telephone directory in order to learn the name that belongs to it. To put it in more general terms, given a set of numbers x, we are searching for a particular number, $x = \tilde{x}$, that has a particular property. To recognize the latter, we have to perform a check. A simple example is a straightforward scheme to find the prime factors of a given integer Z. One will simply check, by division, whether the primes $2, 3, 5, \ldots \leq \sqrt{Z}$ are factors. Assuming for simplicity that Z has only two prime factors, we are looking just for the smaller one. The crucial point is that we will easily recognize it when we are testing it. But it is very hard to calculate it. This is the kind of problem we have in mind.

We assume that a (rather simple) mathematical operation is known that allows us, by trial, to single out a particular number $\tilde{x}$ from a given set of integers $0, 1, 2, \ldots, N - 1$. It is obvious that searching through this set is very time consuming: on average, it will require $N/2$ steps. The question is whether a quantum computer might speed up this procedure. In fact, Grover (1997) devised a quantum algorithm that uses only about $\sqrt{N}$ steps. So it is significantly superior to the classical method.

Let me briefly sketch Grover's algorithm. What we first need is an implementation of the mathematical operation that selects the particular number $\tilde{x}$ we are looking for. This implementation will be the realization, through a network of logic gates, of a unitary transformation U. The latter can be constructed such that it reproduces the state $|x\rangle$ for $x \neq \tilde{x}$ and changes the sign of $|x\rangle$ for $x = \tilde{x}$. Now, as in Shor's period finding algorithm, we benefit from quantum parallelism. Again the first step is to bring the basis states into an equally weighted superposition

$$|\Psi_1\rangle = \frac{1}{\sqrt{N}} \sum_{x=0}^{N-1} |x\rangle \tag{10.17}$$

(cf. (10.11)).

Next the following sequence S is applied to the state (10.17): the unitary operation U, a Fourier transformation, a change of sign of all states $|x\rangle$ except $|0\rangle$, and finally a Fourier back transformation. The result is a slight enhancement of the amplitude of the particular state, at the cost of the amplitudes of the other states. Those amplitudes are equal and hence they are readily calculated from the normalization condition. It was shown by Grover that the amplitude of $|\tilde{x}\rangle$ can be written in the form $\sin \Theta$. The initial value of Θ, Θ_0, is given by $\sin \Theta_0 = 1/\sqrt{N}$, and this value increases by an amount $\Delta\Theta$ that is determined by $\sin \Delta\Theta = 2\sqrt{N-1}/N$. The same enhancement of Θ takes place whenever we apply S. Hence through repeated application we make Θ gradually larger, and after $m \approx (\pi/4)\sqrt{N}$ repetitions we will end up at a value of Θ that comes very close to $\pi/2$ so that the amplitude of $|\tilde{x}\rangle$, $\sin \Theta$, reaches virtually unity. Now it is time to make an energy measurement that will give us the value $\tilde{x}$ we are looking for, with almost certainty. So Grover's algorithm works nicely and, what is the main point, very efficiently.

Care must be taken, however, to avoid overdoing things. In fact, when S is applied too often so that m distinctly exceeds the mentioned optimum value, $\sin \Theta$ will decrease again, which reduces the reliability of the measurement.

References

Anderson, M. H., J. R. Ensher, M. R. Matthews, C. E. Wieman and E. A. Cornell. 1995. *Science* **269**, 198.

Arthurs, E. and J. L. Kelly, Jr. 1965. *Bell. Syst. Tech. J.* **44**, 725.

Aspect, A., P. Grangier and G. Roger. 1982. *Phys. Rev. Lett.* **49**, 91.

Balzer, C., R. Huesmann, W. Neuhauser and P. E. Toschek. 2000. *Opt. Commun.* **180**, 115.

Banaszek, K. and K. Wódkiewicz. 1996. *Phys. Rev. Lett.* **76**, 4344.

Bardeen, J., L. N. Cooper and J. R. Schrieffer. 1957. *Phys. Rev.* **108**, 1175.

Barrett, M. D., J. Chiaverini, T. Schaetz *et al.* 2004. *Nature* **429**, 737.

Bell, J. S. 1964. *Physics* **1**, 195.

Bennett, C. H., G. Brassard, C. Crépeau *et al.* 1993. *Phys. Rev. Lett.* **70**, 1895.

Bohr, N. 1949. Discussion with Einstein on epistemological problems in atomic physics. In P. A. Schilpp (ed.), *Albert Einstein: Philosopher – Scientist*, p. 199. The Library of Living Philosophers, 7. Evanston, Ill.: The Library of Living Philosophers.

Bouwmeester, D., J.-W. Pan, K. Mattle *et al.* 1997. *Nature* **390**, 575.

Brown, R. H. and R. Q. Twiss. 1956. *Nature* **178**, 1046.

Brune, M., E. Hagley, J. Dreyer *et al.* 1996. *Phys. Rev. Lett.* **77**, 4887.

Cirac, J. I. and P. Zoller. 1995. *Phys. Rev. Lett.* **74**, 4091.

Clauser, J. F. and A. Shimony. 1978. *Rept. Prog. Phys.* **41**, 1881.

Davis, K. B., M.-O. Mewes, M. R. Andrews *et al.* 1995. *Phys. Rev. Lett.* **75**, 3969.

Dawydow, A. S. 1978. *Quantenmechanik*. Berlin: VEB Deutscher Verlag der Wissenschaften.

Diedrich, F. and H. Walther, 1987. *Phys. Rev. Lett.* **58**, 203.

Dirac, P. A. M. 1958. *The Principles of Quantum Mechanics*, 4th edn. p. 9. London: Oxford University Press.

Dürr, S., T. Nonn and G. Rempe. 1998. *Nature* **395**, 33.

Einstein, A. 1905. *Ann. Physik* **17**, 132.

Einstein, A., B. Podolsky and N. Rosen. 1935. *Phys. Rev.* **47**, 777.

Franson, J. D. 1989. *Phys. Rev. Lett.* **62**, 2205.

Gerchberg, R. W. and W. O. Saxton. 1972. *Optik* **35**, 237.

Ghosh, R. and L. Mandel. 1987. *Phys. Rev. Lett.* **59**, 1903.

Ginzburg, V. L. and L. D. Landau. 1950. *Zh. Eksp. Teor. Fiz.* **20**, 1064.

Glauber, R. J. 1965. Optical coherence and photon statistics. In C. De Witt, A. Blandin and C. Cohen-Tannoudji (eds.), *Quantum Optics and Electronics*. New York: Gordon and Breach.

Gordon, J. P., H. J. Zeiger and C. H. Townes. 1954. *Phys. Rev.* **95**, 282.

Grover, L. K. 1997. *Phys. Rev. Lett.* **79**, 325.

Hong, C. K., Z. Y. Ou and L. Mandel. 1987. *Phys. Rev. Lett.* **59**, 2044.

Itano, W. M., D. J. Heinzen, J. J. Bollinger and D. J. Wineland. 1990. *Phys. Rev. A* **41**, 2295.

Josephson, B. D. 1974. *Rev. Mod. Phys.* **46**, 251.

Kittel, C. 1986. *Introduction to Solid State Physics*. 6th edn. New York: Wiley.

Kwiat, P. G., A. M. Steinberg and R. Y. Chiao. 1993. *Phys. Rev. A* **47**, R 2472.

Leonhardt, U. 1997. *Measuring the Quantum State of Light*. Cambridge: Cambridge University Press.

Leonhardt, U. and H. Paul. 1993. *Phys. Rev. A* **48**, 4589.

Mandel, L. and E. Wolf. 1995. *Optical Coherence and Quantum Optics*. Cambridge: Cambridge University Press.

Paul, H. 2004. *Introduction to Quantum Optics. From Light Quanta to Quantum Teleportation*. Cambridge: Cambridge University Press.

Pauli, W. 1933. Die allgemeinen Prinzipien der Wellenmechanik. In H. Geiger and K. Scheel (eds.), *Handbuch der Physik* 24/1, 2nd edn. Berlin: Springer Verlag. English translation: Pauli, W. 1980. *General Principles of Quantum Mechanics*. Berlin: Springer Verlag.

Planck, M. 1899. *Sitzungsber. Preuß. Akad. Wiss.*, p. 440. Reprinted in Planck, M. 1958. *Physikalische Abhandlungen und Vorträge*, vol. 1, p. 560. Braunschweig: Vieweg.

Renninger, M. 1960. *Ztschr. Phys.* **158**, 417.

Riebe, M., H. Häffner, C. F. Roos *et al.* 2004. *Nature* **429**, 734.

Roos, C. F., G. P. T. Lancaster and M. Riebe 2004. *Phys. Rev. Lett.* **92**, 220402.

Sauter, T., R. Blatt, W. Neuhauser and P. E. Toschek. 1986. *Opt. Commun.* **60**, 287.

Schleich, W. P. 2001. *Quantum Optics in Phase Space*. Berlin: Wiley-VCH.

Schleich, W., A. Bandilla and H. Paul. 1992. *Phys. Rev. A* **45**, 6652.

Schmidt-Kaler, F., H. Häffner, M. Riebe *et al.* 2003. *Nature* **422**, 408.

Schrödinger, E. 1935. *Naturwissenschaften* **23**, 807, 823, 844. English translation: *Proc. Am. Phil. Soc.* **124**, 323 (1980).

Shor, P. W. 1994. *Proc. 35th Annual Symposium on Foundations of Computer Science*. Los Alamitos: IEEE Press.

Slusher, R. E., L. W. Hollberg, B. Yurke, J. C. Mertz and J. F. Valley. 1985. *Phys. Rev. Lett.* **55**, 2409.

Smithey, D. T., M. Beck and M. G. Raymer. 1993. *Phys. Rev. Lett.* **70**, 1244.

Steane, A. M. 1998. *Rept. Prog. Phys.* **61**, 117.

Taylor, G. I. 1909. *Proc. Camb. Phil. Soc.* **15**, 114.

von Klitzing, K., G. Dorda and M. Pepper. 1980. *Phys. Rev. Lett.* **45**, 494.

von Neumann, J. 1932. *Mathematische Grundlagen der Quantenmechanik*. Berlin: Springer. English translation: von Neumann, J. 1955. *Mathematical Foundations of Quantum Mechanics*. Princeton: Princeton University Press.

Walker, N. G. and J. E. Carroll. 1984. *Electron. Lett.* **20**, 981.

Wootters, W. K. and W. H. Zurek. 1982. *Nature* **299**, 802.

Index